RECREATIONAL WATER QUALITY MANAGEMENT
Volume 1: Coastal Waters

ELLIS HORWOOD SERIES IN
ENVIRONMENTAL MANAGEMENT, SCIENCE AND TECHNOLOGY
Series Editor: Dr YUSAF SAMIULLAH, Principal Environmental Specialist, ACER Environmental, Daresbury, Cheshire
Consultant Editor: Professor R. S. SCORER, Emeritus Professor of Environmental Science, Imperial College of Science and Technology

Bache and Johnstone	MICROCLIMATE AND SPRAY DISPERSION
Bigg	CLIMATIC CHANGES: The Effects of Ocean Processes
Cluckie and Collier	HYDROLOGICAL APPLICATIONS OF WEATHER RADAR
Currie and Pepper	WATER AND THE ENVIRONMENT
Excell and Excell	SAND AND DUST STORMS
Haigh and James	WATER AND ENVIRONMENTAL MANAGEMENT: Design and Construction of Works
Howells	ACID RAIN AND ACID WATERS
James and Eden	ENGINEERING FOR PUBLIC HEALTH
Kay	RECREATIONAL WATER QUALITY MANAGEMENT: Vol. 1 Coastal Waters
Kay and Hanbury	RECREATIONAL WATER QUALITY MANAGEMENT: Fresh Waters
Kovacs, Podani, Tuba, Turcsanyi, Csintalan and Menks	BIOLOGICAL INDICATORS IN ENVIRONMENTAL PROTECTION
Price-Budgen	USING METEOROLOGICAL INFORMATION AND PRODUCTS
Priede and Swift	WILDLIFE TELEMETRY: Remote Monitoring and Tracking of Animals
Ravera	TERRESTRIAL AND AQUATIC ECOSYSTEMS: Perturbation and Recovery
Scorer	CLOUD INVESTIGATION BY SATELLITE
Scorer	SATELLITE AS MICROSCOPE
Scorer	METEOROLOGY OF AIR POLLUTION: Implications for the Environment and its Future
Simpson	GRAVITY CURRENTS: In the Environment and the Laboratory
Simpson	SEA BREEZE AND LOCAL WINDS
Szabo	RADIOECOLOGY AND ENVIRONMENTAL PROTECTION
White, Bellinger, Saul, Hendry and Bristow	URBAN WATERSIDE REGENERATION: Problems and Prospects

RECREATIONAL WATER QUALITY MANAGEMENT
Volume 1: Coastal Waters

Editor:
DAVID KAY
Centre for Research into Environment and Health,
University of Wales

ELLIS HORWOOD
NEW YORK   LONDON   TORONTO   SYDNEY   TOKYO   SINGAPORE

First published in 1992 by
ELLIS HORWOOD LIMITED
Market Cross House, Cooper Street,
Chichester, West Sussex, PO19 1EB, England

 A division of
Simon & Schuster International Group
A Paramount Communications Company

© Ellis Horwood Limited, 1992

All rights reserved. No part of this publication may be reproduced, stored in a retrieval system, or transmitted, in any form, or by any means, electronic, mechanical, photocopying, recording or otherwise, without the prior permission, in writing, of the publisher

Printed and bound in Great Britain
by Bookcraft, Midsomer Norton

British Library Cataloguing in Publication Data

A catalogue record for this book is available from the British Library

ISBN 0-13-770025-3

Library of Congress Cataloging-in-Publication Data

Available from the publisher

Table of Contents

List of Tables	7
List of Figures	9
Preface	13

Chapter 1   15
   Bathing waters : recreation and management
      A Foreword by Sir Hugh Rossi MP
      Chairman of the House of Commons Environment Committee

Chapter 2   19
   The Bathing Water Directive : A perspective from the EC
      Gilles Vincent
      Directorate General XI Commission of the European Communities,
      Rue de la Loi, Brussels

Chapter 3   25
   The role of the NRA in implementing the Bathing Water Directive
      Rupert Grantham
      Topic Commissioner for Microbiology, National Rivers Authority,
      South West Region

Chapter 4   33
   Bathing Water Quality : a local authority perspective
      Stephen Oldridge
      Environmental Health Department, Great Yarmouth Borough Council

Chapter 5   49
   Legal Issues Concerning Bathing Waters
      William Howarth
      Director of the Centre for Law in Rural Areas and Senior Lecturer in
      Law, University College of Wales, Aberystwyth

Chapter 6   71
   Microbiological Aspects and Possible Health Risks of Recreational Water
   Pollution.
      Lorna Fewtrell and Frank Jones
      Research Fellow and Director; Centre for Research into Environment
      and Health, University of Wales, Lampeter SA48 7ED

Chapter 7   89
   Recreational waters : a health risk?
      Rodney Cartwright,
      Public Health Laboratory, St Luke's Hospital, Guildford GU1 3NT

Chapter 8   105
   Statistical aspects of microbial populations in recreational waters
      Edmund B. Pike
      Water Research Centre, Medmenham, UK

Chapter 9   113
   US federal bacteriological water quality standards : a re-analysis
      Jay M. Fleisher
      State University of New York, Brooklyn, NY and Senior Research
      Fellow, CREH, University of Wales, Lampeter

Chapter 10   129
   Recent epidemiological research leading to standards
      David Kay and Mark Wyer
      Director and Research Fellow; Centre for Research into Environment
      and Health, University of Wales, Lampeter SA48 7ED

Chapter 11   157
   Terminal disinfection of wastewater with continuous microfiltration
      Vincent P. Olivieri and George A. Willinghan III
      Memtec America Inc., Timonium, Maryland USA

Chapter 12   175
   UV disinfection : an overview of processes and applications
      William L. Cairns
      Research Manager, Trojan Technologies Inc., 845 Consortium
      151 Court, London, Ontario, Canada N6E 2S8

Chapter 13   201
   Summary and conclusions
      Frank Jones and David Kay
      Centre for Research into Environment and Health, University of
      Wales, Lampeter SA48 7ED

Alphabetical Subject Index   211

List of Tables

Table 6.1	Pathogens from UK recreational water.	72
Table 6.2	Human viruses potentially present in recreational water. (adapted from Rao and Melnick, 1986).	78
Table 7.1	Criteria to be used in assessing causality between environmental exposure and disease (Bradford Hill 1965).	95
Table 8.1	ANOVA for triplicate counts of thermo-tolerant coliform bacteria at two beach sites, 40m apart, sampled at 20-minute intervals on seven occasions.	111
Table 9.1	Data used in regression analysis conducted by the EPA study.	117
Table 9.2	Linear regression of the $\log_{10}$ enterococci density on the rate ratios for total and highly credible GI symptoms. Analysis corresponds to the EPA's original analysis (see text for further details).	119
Table 9.3	Linear regression of the $\log_{10}$ enterococci density on the rate ratios for highly credible GI symptoms. Regression shown below is a re-analysis of the EPA's original analysis without the exclusion of 3 data points (see text for further explanation of this analysis).	119
Table 9.4	Logistic regression of the log odds of gastroenteritis among swimmers (see text for further details of the analysis).	121
Table 9.5	Probability of gastroenteritis to an individual swimmer at the New York City, Boston, and Lake Ponchartrain study locations at 1, 2, and 3 times maximum geometric mean enterococci densities as put forth by current US Environmental Protection Agency guidelines for marine recreational waters (see text for further details).	125
Table 10.1	Recreational water quality standards.	134
Table 10.2	Criteria for indicator for bacteriological densities (USEPA 1986).	138
Table 10.3	Summary results of the Cabelli style and other epidemiological perception studies.	140
Table 10.4	Summary statistics, all data, Langland Bay 2/9/89.	142
Table 10.5	Crude perceived symptom attack rates per 1,000 for the a) bather cohort and b) non-bather cohort.	144
Table 10.6	Significance values (p) for Chi^2 analysis of significance between bather and non-bather perceived symptom attack rates.	145

List of Tables

Table 10.7 Crude clinical symptom attack rates per 1,000 for ear and throat swabs. — 146

Table 10.8 Significance values (p) for Chi^2 analysis of significance between bather and non-bather clinical symptom attack rates. — 146

Table 10.9 Faecal sample results, number of positive occurrences. — 147

Table 11.1 The level of suspended solids (SS), biochemical oxygen demand (BOD) and faecal coliform before and after CMF obtained during the commissioning in the fall of 1990 of the 4 megalitre plant at Blackheath, New South Wales, Australia. — 160

Table 11.2 Relative microbial balance from raw wastewater for the selected microorganisms at three plants in the United States. TC = total coliform, FC = faecal coliform, FS = faecal streptococci, ENT = enterococci, C.p. = *Clostridium perfringens*, Sal. = *Salmonella* spp., MSBV = male-specfic bacterial virus and HEV = human enterovirus. — 169

Table 12.1 Comparative sensitivity of microbes to UV disinfection. — 181

Table 12.2 Mutation rates for faecal coliform organisms. — 183

Table 12.3 Capital cost for plant of different sizes. — 197

List of Figures

Figure 8.1	Relationship between counts of colonies and volumes of activated sludge mixed liquor plated by spreading on casitone-glycerol-yeast extract agar; incubation at 22°C for 6 days (DoE 1972). The line of equality and numbers of overlapping points are shown.	109
Figure 9.1	Regression lines depicting the association between the rate difference of GI symptoms among swimmers and non-swimmers and enterococci density as reported by the EPA study. From Cabelli *et al*. (1982), Figure 2, Page 614.	117
Figure 9.2	Regression lines depicting the association between the rate of ratio of GI symptons (swimmers/non-swimmers) and enterococci densities as reported by the EPA study. From Cabelli *et al*. (1982), Figure 3, Page 615.	118
Figure 9.3	Probability of gastroenteritis among swimmers versus enterococci density.	123
Figure 10.1	Bathing water quality in England and Wales 1990.	130
Figure 10.2	Dose-response relationships produced by the work of Cabelli *et al*. (1982)	136
Figure 11.1	Log levels of natural populations of *Salmonella* in undisinfected wastewater effluent at three plants in the United States.	158
Figure 11.2	Log levels of male-specific bacterial virus (host Salmonella typhimurium WG49) in undisinfected wastewater effluent at two plants in the United States.	158
Figure 11.3	Plan view of the 4 megalitre CMF at Blackheath in Australia.	160
Figure 11.4a	Levels of suspended solids (SS) before and after CMF filtration in extended aeration effluent at the Round Corner pilot plant in Australia.	162
Figure 11.4b	Levels of biochemical oxygen demand (BOD) before and after CMF filtration of extended aeration effluent at the Round Corner pilot plant in Australia.	162
Figure 11.5a	Log levels of indicator bacteria before and after CMF filtration of extended aeration effluent at the Round Corner pilot plant in Australia.	163

List of Figures

Figure 11.5b Log levels of human enterovirus before and after CMF filtration of extended aeration effluent at the Round Corner pilot plant in Australia. 163

Figure 11.6 CMF flow (gallons per day per square foot of membrane area) and trans membrane pressure (TMP) for one filter per cycle for dead end operation at the Back River WWTP in the United States. 164

Figure 11.7a Levels of suspended solids (SS) before and after CMF filtration of activated sludge effluent at the Back River pilot plant in the United States. 165

Figure 11.7b Levels of biochemical oxygen demand (BOD) before and after CMF filtration of activated sludge effluent at the Back River pilot plant in the United States. 165

Figure 11.8a Log levels of indicator bacteria before and after CMF filtration of activated sludge effluent at the Back River pilot plant in the United States. 166

Figure 11.8b Log levels of male-specific bacterial virus before and after CMF filtration of activated sludge effluent at the Back River pilot plant in the United States. 166

Figure 11.9a Levels of suspended solids (SS) before and after the Membio process during trials at Malabar in Australia. 167

Figure 11.9b Levels of biochemical oxygen demand (BOD) before and after the Membio process during trials at Malabar in Australia. 167

Figure 11.10 Log levels of faecal coliform before and after the Membio process during trials at Malabar in Australia. 168

Figure 11.11a Levels of suspended solids (SS) before and after the Membio process at a pilot plant in the United Kingdom. 170

Figure 11.11b Levels of biochemical oxygen demand (BOD) before and after the Membio process at a pilot plant in the United Kingdom. 170

Figure 11.12a Log level of indicator bacteria before and after the Membio process at a pilot plant in the United Kingdom. 171

List of Figures

Figure 11.12b	Log level of bacterial virus before and after the Membio process at a pilot plant in the United Kingdom.	171
Figure 12.1	Electromagnetic Spectrum with Expanded Scale of Ultraviolet Radiation. (1 nanometre = 10^{-9} metre.)	176
Figure 12.2	Similarity between the inactivation of E. Coli cells and the absorption spectrum of nucleic acids. (from Harm, W. "Biological Effects of Ultraviolet Radiation" Pg. 29, Cambridge University Press, Cambridge, 1980).	177
Figure 12.3	Examples of DNA photoproducts formed in UV-irradiated cells. (from Harm, W. "Biological Effects of Ultraviolet Radiation' Pg. 29, Cambridge University Press, Cambridge, 1980).	179
Figure 12.4	Demonstration of relative effectiveness of UV: (1) *Escherichia coli* B (2) *Salmonella typhosa* (3) *Staphylococcus aureus* (4) Polio type 1 virus (5) Coxsackie AZ virus (6) Adenovirus Type 3.	180
Figure 12.5	Counts of UV irradiated and non-irradiated *E. Coli* (NAR) in Big Otter Creek at different times after discharge to the receiving water.	182
Figure 12.6	Fraction survival curve of UV irradiated faecal coliforms from wastewater.	187
Figure 12.7	Relationship between wastewater disinfectability and the level of suspended solids.	189
Figure 12.8	Retrofit of UV into an existing chlorine contact tank at Northwest Bergen County's wastewater plant.	195
Figure 12.9	Equipment layout showing modules (racks of lamps) placed in parallel and series.	196

PREFACE

Recreational water quality is an environmental issue of great importance. DoE research suggests that the United Kingdom public are more concerned about sewage pollution of their beaches than even their drinking water quality. Only the chemical contamination of rivers having a higher profile than bathing water quality in the public consciousness. The political response to this public concern has resulted in recent prodigious efforts to monitor bathing water quality in accordance with the Bathing Water Directive (76/160/EEC). Since the early 1980s the United Kingdom has increased its EC monitoring sites from a paltry 27 to 446 in 1990. In addition to the expanded monitoring, great strides have been taken to make the resultant data available and understandable to the lay public through the provision of display boards at beaches. Some local environmental health departments (e.g. Southend) had been doing this for many years, however, the DoE guidance, issued in 1990, has introduced national coordination in the dissemination of environmental information which has not been seen in many other fields. Public awareness has reinforced the demand for further information and significant improvement in recreational water quality. Such improvement can only come about as a result of capital intensive sewage treatment and disposal schemes. The national (UK) commitment to effect such improvement programmes and bring its coastal recreational waters to the mandatory level specified by the European Commission will cost several billion pounds and accounts for a significant proportion of the increases in consumer bills now being levied by the privatised water companies.

Given this frenetic activity and huge financial commitment, the proverbial visitor from another planet might be excused for concluding that polluted bathing waters produced a significant and proven health risk and that this focusing of the national will resulted from a desire to stamp out some disease epidemic related to recreational water exposure. In fact this is not the case. In the modern period little hard data exist to prove that serious diseases have been contracted from marine recreational waters. There is a growing body of knowledge which indicates that minor gastrointestinal, upper respiratory and skin complaints are contracted from bathing in sewage contaminated recreational waters. It is these possible illnesses that standards for recreational waters and compliance assessment programmes seek to limit. Herein lies the dilemma for public health and water management agencies who have no robust scientific information on which to base standards or with which to assure the public that no health risk is evident at sites which PASS current standards in force.

This book addresses these problems from the United Kingdom's preoccupation with this issue resulting the 1990 Parliamentary enquiry by the Commons Environment Committee. This investigation produced one the most incisive and balanced examinations of this issue available. The context of this enquiry is explained in his Foreword by the Chairman of that Committee, Sir Hugh Rossi (Chapter 1). The role of international and national control agencies is then presented by Gilles Vincent of the European Commission (Chapter 2) and Rupert Grantham of the UK National Rivers Authority (Chapter 3) and Steve Oldridge of Great Yarmouth BC (Chapter

4). Each have operational responsibility for the application of the current Directive (76/160/EEC) and are at the sharp end of implementation within agencies of contrasting scale and responsibilities. These policy actions and outcomes are played against a backcloth of legal structure and microbiological hazard. The next two chapters explore these aspects with contributions from Bill Howarth (Chapter 5). Lorna Fewtrell and Frank Jones (Chapter 6) provide an authoritative review of the microbiological hazards experienced in the coastal zone and Rodney Cartwright (Chapter 7) assesses the public health impact of the pathogens present. Relating this to policy and standards development requires significant examination of the statistical distributions of microbial populations and this is addressed by Edmund Pike (Chapter 8) who with his WRc colleagues has researched this field for over a quarter of a century. Jay Fleisher (Chapter 9) then provides a cautionary tail centred on the USEPA approach to standards development, where recent examinations of the original epidemiological studies conducted in the 1970s have thrown some doubt on the validity of the USEPA recreational water quality standards published by USEPA in 1986. This theme is maintained by Kay and Wyer (Chapter 10) who demonstrate the poor epidemiological foundation for most international standard systems and suggest a novel research protocol which might overcome some of the flaws in the USEPA design identified by Fleisher (Chapter 9). Appropriate treatment for coastal discharges is clearly important given the huge cost implications and heightened public awareness. The final two chapters centre on new and exiting treatment options for coastal discharges both of which claim little or no by-product formation. Olivieri and Willinghan (Chapter 11) concentrate on microfiltration, which is under full scale plant trials in the UK and Australia whilst Cairns evaluates the role of UV disinfection for coastal discharges (Chapter 12). In the final Chapter (Chapter 13) Frank Jones and David Kay summarise current policy options and the difficult choices facing environmental decision makers.

David Kay

ACKNOWLEDGEMENTS

Most of the papers contained in this volume derive from a conference organised in December 1990 for the Centre for Research into Environment and Health by Oxford Conferences Limited. We are most grateful to Mrs Cathy Pownall of Oxford Conferences for her professional and smooth organisation of this conference. The manuscript was prepared by Ms Paula Hopkins and Mr Jerome Whittingham of CREH. Dr Mark Wyer and Mr Trevor Harris prepared the figures and completed all of the DTP. For all these inputs we are most grateful.

The sad and untimely death of Vincent Olivieri occurred during production of this book. He will be greatly missed by all workers in the field of recreational water quality. His great contribution to the subject will remain for many years.

Chapter 1

BATHING WATERS : RECREATION AND MANAGEMENT

A Foreword by Sir Hugh Rossi MP
Chairman of the House of Commons Environment Committee

BACKGROUND TO THE CURRENT SITUATIONS

From Victorian times, the preferred method of sewage disposal from coastal towns was to discharge it untreated through sea-outfalls. The original engineers of our sewage systems opted for short outfalls, often discharging directly on to the beaches. A lack of investment in upgrading coastal outfalls meant that many beaches continued to be polluted by sewage, so that in 1986, when the first national survey was undertaken, over one-third of Britain's bathing waters failed to meet the microbiological standards of the EC Bathing Water Directive.

In order to clean up the beaches, Britain embarked on a major construction programme of replacing the old short outfalls with long outfalls. These still discharge sewage which is untreated, but at a point 3-5 km from the shore. Previously, the solid material has been macerated and screened to remove large solids and plastics before the sewage leaves the headworks. The action of sea and sunlight then breaks down the sewage.

The water industry held that well-designed long sea-outfalls are a cost-effective way of using the natural treatment capacity of the coastal seas to deal with sewage. Others have questioned this reliance on the "dilute and disperse" principle which has been discredited in other areas of pollution control, arguing that the long outfall could be to the oceans what the tall chimney had proved to be for the atmosphere.

We began our own inquiry amid mounting public concern over possible health risks associated with sewage discharges to the sea. Research in the US (by Victor Cabelli) indicated that bathing in seawater (of a quality which would not normally be considered as polluted) could cause ear, nose and throat infections and tummy upsets. In the summer of 1989, the Department of the Environment began its own research into the health risks of sea-bathing on a beach which passed the EC Directive - the Langland Bay Study; and it was published while our inquiry was in progress. The study succeeded in establishing a valid methodology for assessing the epidemiology, but concluded that greatly extended studies would be required to obtain statistically-significant results.

CONCERNS EXPRESSED TO THE COMMITTEE

The committee received a large amount of evidence critical of Government policy on sewage treatment from a variety of different perspectives. A number of people living near polluted beaches complained about the health and visual impacts of raw sewage. Environmental groups in areas where long sea outfalls were planned expressed doubt about whether a long sea outfall would be effective and campaigned for land-based treatment. Windsurfers and sailing clubs complained about sewage in coastal waters particularly from long sea outfalls which is difficult for water-sports enthusiasts to avoid. A number of professionals working in the water industry complained about defects in the EC Bathing Water Directive particularly with regard to salmonella and enterovirus standards. Since compliance with the EC Directive has been the goal of British environmental policy, this raises serious questions.

MUNICIPAL WASTE WATER TREATMENT DIRECTIVE

In the course of our inquiry, the European Commission published the draft Directive on Municipal Waste Water, which sets out minimum standards for the treatment of sewage before discharge to rivers and coastal waters. In effect, it would make secondary treatment the norm for sewage works serving a population equivalent to more than 10,000 people, so that, before discharge the sewage would have to be held in sedimentation tanks after maceration and screening to settle out grit, sand and similar fine solids (i.e. primary treatment) and then subjected to biological oxidation so as to reduce the biological oxygen demand (BOD) of the resultant effluent (i.e. secondary treatment). For discharges to open waters with strong currents, primary treatment may suffice; for particularly sheltered waters or estuaries, further treatment beyond the secondary stage may be required.

On the eve of the Third North Sea Conference at the Hague, the Secretary of State for the Environment announced that he was accepting the terms of the draft Directive, and cited the Committee's inquiry as a factor in the Government's decision. At the same time, the Minister of Agriculture, Fisheries and Food announced that the practice of dumping sewage sludge at sea would be terminated at the end of 1998. Both of these decisions were warmly welcomed in the Environment Committee's Report; however, they have important consequences both for the water industry and for the consumer.

THE PROBLEM OF SLUDGE

At the moment, about 100,000 tonnes of dry sludge is disposed of inland annually; about half of it goes to farmland, 20 per cent to landfill, and five per cent is incinerated. About 230,000 tonnes are dumped at sea. The new policy will mean finding a new disposal route not only for the sludge already dumped at sea, but also for the considerable quantities of additional sludge which will be produced by the primary and secondary treatment of the sewage which is currently disposed of through outfalls after maceration and screening alone. Britain is still faced with the prospect of legal action by the European Commission over our failure to clean-up beaches by the Directive's 1985 target. The programme of clean-up has switched from long sea outfalls to treatment works and the Government is aiming to bring all bathing waters into line by 1997. It remains to be seen whether Brussels will withdraw its legal action.

It is clear that higher environmental standards will entail higher costs for the water industry and, ultimately, for the consumer. The Minister of State, Mr Trippier, estimated the impact of the new policy on the industry's investment programme at about £1.5 billion, or an extra five to six per cent, on average, over ten years, with variations depending on local conditions. This point was not lost on the Director General of Water Services (OFWAT), who criticised the Secretary of State's announcement very soon after it was made and, subsequently, in his first Annual Report, expressed concern about its likely effect on customers' bills. Lord Crickhowell, Chairman of the NRA has, in return, criticised OFWAT for going beyond its powers in a way which jeopardises urgent environmental improvements.

If nothing else, the events of 1990 have demonstrated that the climate of public opinion on environmental issues has changed dramatically and that Government is having to respond to this new mood. But the sewage disposal story also points up another issue - that while it is crucial that we tackle environmental pollution, there is rarely a simple *technical fix* which will solve the immediate problem without creating others in its wake. The decision massively to reduce the pollution load in our coastal waters is undoubtedly right; but people are going to have to learn to live with consequences which they may not have originally foreseen.

Chapter 2

THE BATHING WATER DIRECTIVE : A PERSPECTIVE FROM THE EC

Gilles Vincent
Directorate General XI Commission of the European Communities,
Rue de la Loi, Brussels

INTRODUCTION

The management of bathing waters is a topic very much in the public eye and so it is particularly important that all concerned, administrators, scientists, academics and, not least, the general public understand the issues clearly. The Bathing Water Directive (EEC, 1976) is the basis for the management of bathing water in the EEC, and the purpose of this paper is to discuss some aspects of the Directive and to give an account of the Commission's part in applying it.

The Directive is a legal instrument, which is ultimately founded on the Treaty of Rome. It therefore imposes binding legal obligations upon Member States and, where necessary, these can be enforced by the European Court of Justice. The judicial nature of the Directive is central to the understanding of it.

The objectives of the Directive are given in its first recital. They are to protect the environment and public health by reducing the pollution of bathing water and by protecting such water against further pollution. The remaining recitals set out why this should be done and provide a reasoned basis for the individual Articles in the Directive. It is to be noted that the recitals are an integral part of the Directive and where necessary they can serve to illuminate or amplify the meaning of individual Articles.

THE DIRECTIVE

The Directive must, of course, be considered as an entity. However, it will be useful to examine certain parts in detail.

The principal obligation, which is contained in Article 4, is that Member States must ensure that the quality of bathing water meets the Directive's standards. Much of the rest of the Directive is concerned with defining this obligation and the means needed to comply with it.

Perhaps the most significant sentence in the Directive is that which comprises Article 6(3). This states:

> *Local investigation of the conditions prevailing upstream in the case of fresh running water, and of the ambient conditions in the case of fresh still water and sea water should be carried out scrupulously and repeated periodically in order to obtain geographical and topographical data and to determine the volume and nature of all polluting and potentially polluting discharges and their effects according to the distance from the bathing area.*

In other words, the competent authorities must inform themselves of the factors affecting water quality. This puts the Directive's monitoring requirements into context. The sampling referred to in Articles 6(1) and 6(2) is not intended to discover the quality of bathing water. This should be known already, and sampling is intended either to confirm that water quality is as expected or, alternatively, to indicate that not all sources of pollution have been identified. Where necessary, investigations must be made to identify such further sources of pollution; and then remedial action should be taken to protect bathing water quality.

The Directive therefore requires Member States to be pro-active in regard to bathing water quality. It is not sufficient just to measure quality. The quality standards for bathing water are set out in the Directive's annex. These relate to a range of parameters which define the microbiological and aesthetic quality of bathing water. The principal microbiological parameters are total coliforms and faecal coliforms. These are indicators of faecal pollution and so of the possible presence of pathogens. Although the evidence they provide is indirect they remain useful parameters. They are readily measured and their significance is widely understood. There are three further microbiological parameters, faecal streptococci, salmonella and enteroviruses. The measurement of these parameters is only required when the competent authority has reason to believe that they are present or when water quality has deteriorated. Mandatory values are set for the following physiochemical parameters;

> pH,
> colour,
> mineral oils,
> surface active substances,
> phenols and
> transparency.

with the exception of pH, all must be measured regularly. These parameters are measures of the aesthetic quality of bathing water.

FUNCTIONS OF THE COMMISSION

It is widely thought that the only Community institution concerned with the Directive is the Commission. This is not quite correct. The Commission proposed the Directive, and did so after taking account of advice it had received from national experts and of information available in the scientific literature. However, the Directive as it now exists was adopted by the Council of Ministers. The Council took account of opinions expressed by the European Parliament and by the Economic and Social Committee and, of course, the views of the individual Member States.

Once the Directive had been adopted, which was on 8 December 1975, it became the Commission's responsibility to ensure that Member States apply it correctly. This duty upon the Commission follows from the EEC Treaty. In this case, it has two formal tasks :

(i) to receive reports on the quality of bathing water submitted by Member States in accordance with Article 13, and to publish this information; and

(ii) where necessary, to take legal proceedings against Member States who appear not to be applying the Directive correctly.

These tasks will be discussed below, but it should be noted that as far as possible the Commission will seek to work in co-operation with Member States, which is by far the best way to achieve the purposes of the Directive. That is not to say that rules and formality are not important. They do have an important part to play, but they should assist, not dominate, the process of implementing the Directive.

REPORTS

The obligation, contained in Article 13, for Member States to report periodically on the quality of their bathing waters allows the Commission to be informed of water quality in the Member States. The need to present reports has also led Member States to identify individual bathing areas and to ensure that bathing water quality is measured according to the Directive's prescriptions. The information received allows the Commission to publish reports which enable Member States and the general public to be informed of bathing water quality throughout the Community.

These reports attract considerable attention, not least from Member States. It is clear that no Member State wishes to be out of line with its partners. This has led to substantial changes in national policies in some countries and to a progressive improvement in bathing water quality throughout the Community. These reports have certainly contributed to the fundamental shift in public opinion which has taken place in recent years and to a continuing demand for high quality bathing water. The effectiveness of public opinion is not to be underestimated.

It should be noted that in publishing the Article 13 reports the Commission is not only complying with the specific obligation contained in the Directive but it is also taking full account of its last recital. This refers to the need for the public to receive objective information on the quality of bathing water.

LEGAL PROCEEDINGS

It sometimes happens that the Commission considers that a Member State is not complying with its obligations under the Directive. The Commission has a duty to ensure that Directives are applied correctly, and if necessary it can refer cases to the Court of Justice of the European Communities. Such a reference would be a matter of last resort, and would only be made when no other option remained.

The four steps leading to proceedings in the Court are, briefly, as follows.

> The <u>first</u> is a formal letter setting out why it appears to the Commission that the Directive is not being applied correctly. In many cases the Member State either provides a satisfactory explanation or takes appropriate action, and the matter is disposed of.

> If, however, the Commission is not satisfied then the <u>second</u> step is for the Commission to send a further letter - often known as an Article 169 letter because it refers to the Article in the EEC Treaty which is the basis for the Commission's future legal proceedings. Again the Member State has the chance to dispose of the matter either by providing a satisfactory explanation or taking appropriate action.

> Should it not do so the <u>third</u> step is for the Commission, the College of Commissioners, to adopt a Reasoned Opinion.

> In effect this sets out the case the Commission would present to the Court of Justice if necessary. It will include a summary of the facts as known by the Commission, the legal arguments and counter-arguments presented by the Commission and by the Member State, a statement of the action taken by the Member State to comply with the Directive, and the reasons why the Commission considers that the Directive is not being complied with The Reasoned Opinion is the last opportunity for the Member State to provide a satisfactory explanation or take appropriate action.

> If it does not, then the <u>fourth</u> step is for the case to come before the Court of Justice.

It should be stressed that it is rare for a case to go all the way to the Court. It is much more usual for matters to be brought to a satisfactory conclusion before that point is reached. However, it is important to note that the Commission has both the power and the duty to refer to the Court cases where, after enquiry, it appears that a Member State is failing to apply a Directive correctly.

The Commission has a further part to play. It has the right to propose legislation to the Council. It follows that it should keep Directives under review, and consider if changes are needed. The Commission must therefore be aware of difficulties in applying existing legislation, Directive 76/160/EEC in this case, and must keep itself informed of the latest scientific knowledge. Information is obtained in several ways:

    formally from Member States;

    from meetings of national experts arranged by the Commission;

    by attendance at scientific meetings;

    from the scientific literature; and

    from individual members of the public and from organised groups who make representations to the Commission.

The following is a selection of points which have been raised recently.

1. *What is a bathing water ?*

2. *Should the term bathing water extend to water used for wind-surfing, sailing and water sports generally ?*

3. *Given that it is essential to protect the health of bathers, what are the best parameters to measure and what are the appropriate numerical standards ?*

4. *Although coliform bacteria are relatively easy to measure, are they the best indicators ?*

5. *What better indicators are there ?*

6. *Is one indicator sufficient to assess the possible risk to the health of bathers ?*

7. *Are there satisfactory methods of disinfection ?*

8. *Is disinfection of sewage useful or does it merely destroy indicator organisms while leaving pathogens intact ?*

9. Is the present sampling regime satisfactory ?

These are questions which merit careful consideration both within the Commission and also with external advisers. This consideration is a continuing process, and the Commission is, in effect, invited to present proposals for a new bathing water Directive. However, a decision to present a new proposal would not be reached lightly. The fundamental need to protect the environment and public health remains, and to be worthwhile any new proposal would need to be demonstrably better than the present Directive. It would also be necessary for a new proposal to command general support within the Council.

This suggests that an evolutionary approach will be best. Fundamental changes will probably not be needed but the Directive must be kept up to date. The Commission has suggested a mechanism to do this. It has invited the Council to set up a regulatory committee with the power to adapt certain Directives, including the Bathing Water Directive. This proposal is at present under consideration by the Council.

The Directive was adopted in 1975. Since then it has led to substantial improvements in bathing water quality and has also helped to raise the level of interest in environmental matters generally. It has been a success, but there is still more to be done.

Member States must play their part and complete the work of bringing the remaining bathing waters up to Community standards. The Commission too must play it's part. Where necessary it must ensure that the Directive is applied correctly. It must also ensure that the Directive is fully up to date. We must build the future on the experience of the past.

REFERENCE

EEC (1976) Council Directive of 8th December 1975 concerning the quality of bathing water. *Official Journal L31*, 1-7.

Chapter 3

THE ROLE OF THE NRA IN IMPLEMENTING THE BATHING WATER DIRECTIVE

Rupert Grantham
Topic Commissioner for Microbiology, National Rivers Authority, South West Region

BACKGROUND TO THE NATIONAL RIVERS AUTHORITY

The National Rivers Authority (NRA) was set up under the 1989 Water Act with a wide range of responsibilities. These are reflected in the Authority's aims for the management of water quality and water resources, the provision of adequate flood defence for people and properties, the maintenance and development of inland fisheries, the conservation of the water environment and the promotion of recreation. Waters for which the NRA has responsibilities include groundwaters, freshwaters, estuaries and coastal waters to three nautical miles off shore. The NRA has statutory duties for management of water quality which are:

a. To conserve and enhance amenity;

b. To monitor the extent of pollution in controlled waters;

c. To regulate discharges to these waters;

d. To achieve water quality objectives;

e. To carry out supportive research.

A key aspect here will be the establishment, by the Secretary of State for the Environment, of a system of Statutory Water Quality Objectives (SQOs) which will apply to all controlled waters. The timetable for establishment and achievement of

SQOs has yet to be confirmed, but it is anticipated that proposals for such a system will be issued for consultation during 1992.

In order that the NRA can fulfil its duties it has been granted wide powers which allow it to prevent pollution from occurring, to prosecute polluters who cause or knowingly permit pollution, to remedy or mitigate pollution and to recover the costs of its operations.

The NRA operates very much in the public domain and is required to make publicly available maps of all controlled waters, registers of water quality objectives, details of consents and the results of its monitoring of discharges and of receiving water quality. In addition, the NRA has statutory duties in relation to the provision of advice and exchange of information on matters of water pollution. These include the requirements to advise the Department of the Environment (DoE) and the Ministry of Agriculture, Fisheries and Food (MAFF) and to exchange information with the water companies.

There are important limitations on the NRA's responsibilities, particularly in that the Authority has no direct responsibility for the quality of drinking water or for matters of public health.

EC BATHING WATER DIRECTIVE APPLIED IN ENGLAND AND WALES

The EC Directive concerning the quality of bathing waters (1975) was an early piece of European environmental legislation which sought to address a widespread and complex problem. The Directive aims to protect the environment and public health. Article 6 of the Directive identifies the role of the competent authority for monitoring and investigation. In 1990, Mr Heathcoat-Amory identified the NRA as the competent authority in England and Wales for the purposes of implementing the EC Bathing Water Directive. The Water Act obliges the NRA to monitor the quality of bathing waters and to achieve the standards laid down by the Directive. Regulations governing this are, at present, in draft form. Ultimately, it will be via SQOs that bathing waters will be protected. Once SQOs have been set there will be a duty on the NRA and the Secretary of State to achieve them at all times. SQOs will be reviewable at five yearly intervals or earlier if requested by the NRA. SQOs for tidal waters will probably be introduced in the mid 1990's and are likely to have three components. First, the achievement of the requirements of EC directives, second the achievement of a target water quality classification and third the achievement of appropriate use related objectives.

The Bathing Waters Directive stipulates compliance with its standards to be achieved by 1985 but allows derogations to be applied for which are justified on the basis of management plans. By 1985 only 27 waters within the United Kingdom had been identified as coming within the scope of the Directive. In response to subsequent pressure from the European Commission (EC), the number of UK identified bathing waters was increased substantially and in 1990 the total stands at 446. Within England and Wales, additions to this list of identified waters may be recommended by the NRA on the basis of the popularity of such waters for bathing.

At present, no inland waters within the United Kingdom have been identified in terms of this Directive. However, the NRA will be reviewing the recreational use of inland waters when advising the Secretary of State on the establishment of appropriate SQOs.

DIRECTIVE STANDARDS

The Directive lays down standards for 13 parameters and scope for Member States to apply standards for a further 6 parameters. Methods of measurement are also specified. For some parameters visual assessment is adequate whereas for others an analytical method or choice of methods are defined. For example coliform levels may be measured either by multiple tube fermentation with a most probable number (MPN) count or by membrane filtration. Results obtained will be very much dependent on the method chosen. In general, EC bathing waters within England and Wales are monitored 20 times during each bathing season. The Directive does however allow for a 50% reduction in the frequency of monitoring where results of water quality obtained in previous years are appreciably better than the standards laid down by the Directive. For a number of parameters, the requirement is that the concentration must be checked when inspection shows that the substances 'may be present' or that water quality has deteriorated. These parameters include faecal streptococci, salmonella and enteroviruses. For this reason, these organisms are not checked on all occasions at all sites. For certain parameters, for example transparency and colour, waivers may be granted that take account of exceptional weather or geographical conditions or the natural enrichment of bathing waters.

The most important standards of the Directive are those laid down for the microbiological parameters. Mandatory standards (I values) must be achieved and guideline standards (G values) are those which member states should endeavour to achieve. Each microbiological standard has a statistic associated with it, this reflects the variable nature of the measured micro biological quality of natural waters. Thus, 95% of samples must comply with the mandatory standards for the coliform parameters, whereas only 80% of samples need comply with the more stringent guideline standards for these organisms.

NRA MONITORING STRATEGY

The variability of microbiological results reflects both a lack of analytical precision and the changeable nature of the waters themselves. It is not practical to monitor the quality of bathing waters continuously and it is recognised that 20 discrete results of bathing water quality obtained during a season represents no more than a series of snapshots of water quality. It is therefore important to view results on a cumulative basis as several years data are required for a meaningful picture of water quality at any one site.

The NRA tests for all parameters specified by the Directive unless waivers have been granted. Inspection at the time of sampling may also include an assessment of the contamination of the beach itself and the environmental conditions prevailing at the time, for example wind, tide and intensity of sunshine. For the purposes of compliance assessment, the NRA's monitoring commences at the beginning of May, two weeks before the start of the bathing season and continues through until the end of September. This equates broadly with weekly sampling. Samples are collected in approximately 1m depth of water 30cm below the surface. In accordance with the requirements of the Directive, samples are taken at places where the daily average density of bathers is highest. However where requirements for improvements have been identified, the NRA requires compliance along the full length of the bathing area. In support of this, additional detailed investigations are undertaken to establish the extent of water quality problems in a given bathing water and to identify the sources of pollutants reaching that bathing water. The NRA also has draft proposals to monitor non-EC bathing waters.

COMPLIANCE ASSESSMENT

The results of monitoring are available on request from NRA regional offices. These results are also provided to maritime district councils, the Blue Flag Campaign organisers and to the Department of the Environment who pass the results on to the European Commission. It is important to recognise that compliance with the requirements of the Directive is ultimately determined by the European Commission. Early in 1991, the NRA expects to produce its first detailed report of bathing water quality in England and Wales giving details of the NRA's approach, the results of monitoring and the background to bathing water problems which have been identified (NRA, 1991a).

IMPROVING THE STANDARDS

The Directive has come in for much criticism in recent years. It should, however, be recognised that when the Directive was issued in 1975, it was helpful in setting standards for easily measurable indicators of pollution. Nevertheless, the bacterial and viral standards of the Directive have little epidemiological or scientific basis. An obvious anomaly in the Directive is that it requires zero enteric viruses to be present in 10 litres of water, yet allows a level of enteric bacteria. It would be impossible to guarantee compliance with this standard for viruses. Article 10 of the Directive set up a Committee on Adaptation to Technical progress. Unfortunately, to date, no initiatives have been forthcoming from this Committee. The standards are in need of updating on the basis of modern epidemiology and science but it is recognised that this will be a slow process. The NRA are keen to ensure that improvements to bathing water quality are meaningful and represent a sound investment for the future. Various approaches have been tried around the world, in deriving water quality standards on the basis of epidemiology. Whilst this is entirely desirable it is also extremely difficult because the health risk of bathing in polluted waters is determined by:

    a. The health of the community served by the local discharge(s);

b. The bathers' resistance to infection;

c. The quantity of water ingested by bathers.

As previously stated, the NRA has no direct responsibility for public health and is unable to advise on public health matters, yet it does have responsibility for managing the quality of recreational waters. This being so, the Authority is keen to ensure that improvements to bathing water quality do provide the level of health protection that bathers may reasonably expect. Water quality standards currently available to the NRA are based on the bathing water Directive. The NRA is now promoting an extensive research and development programme which includes projects aimed at deriving more suitable standards. The Authority is co-funding an epidemiological study of UK coastal bathing waters, due for completion at the end of 1992, and separately is researching the derivation of standards for water contact sports for both inland and coastal waters. Other microbiological R & D initiatives include a project to identify the behaviour of micro-organisms in surface waters, a project to define and improve standard methods of analysis and a project appraising the efficacy and environmental impact of sewage disinfection techniques.

SEWAGE DISINFECTION

Disinfection of sewage is an attractive option where traditional 'treatment at sea' is either impossible or impractical. A sewage disinfection process will be acceptable to the NRA, provided it;

a. Reduces levels of indicator organisms;

b. Reduces levels of pathogenic organisms; and

c. Does not result in adverse environmental side effects.

As yet, no disinfection process has been proven in relation to the above three criteria. The NRA is therefore carrying out an extensive programme of field and laboratory studies to examine the effectiveness and environmental impact of those processes which are of most interest to the water companies. Field studies are generally carried out in collaboration with water companies with all trials meshing into a national programme of information gathering.

ACHIEVING THE DIRECTIVE STANDARDS

The NRA has a duty to comply with all standards laid down by the Directive. Standards for pH, colour, mineral oils and tarry residues may be influenced by industrial discharges, however these are not a source of major concern in relation to bathing water. A few localised problems do exist, for example in relation to the coal industry, oil spills, the disposal of dredge spoil, mining activities and food processing industries. Regarding the microbial standards, the NRA's principal concern is with the direct discharges of sewage to the sea. Also widespread problems result from rivers carrying effluent from inland sewage treatment works, stormwater discharges, agricultural and industrial waste and diffuse inputs from farms and soakaways. A vast majority of rivers carry concentrations of bacteria higher than those specified by the mandatory standards of the Directive.

Effective control of these polluting sources may be achieved by improving the quality of discharges, relocating discharges, increasing sewer capacity to reduce overflow frequency and, in theory, establishing protection zones to reduce diffuse inputs.

The flow and composition of sewage varies considerably. On average, more than 99% is water with the remainder being principally degradable organic matter plus some persistent contaminants such as metals, pesticides and debris. 'Preliminary' sewage treatment removes gross solids, grit and plastics, commonly using fine screens with apertures up to 6mm. 'Primary' sewage treatment removes the remaining inorganic solids and much of the suspended organic solids. 'Secondary' sewage treatment follows primary treatment and comprises biological filtration or aeration to reduce the oxygen demand of effluent and remove most remaining solids. 'Tertiary' treatment includes various polishing processes, notably nutrient removal and disinfection.

In anticipation of the forthcoming EC Urban Waste Water Directive and as a prelude to the next North Sea Conference, in March of this year (1990) the Secretary of State announced that all sewage discharges of more than 1500 cubic metres per day would receive secondary treatment if discharging to an estuary or primary treatment if discharging to coastal waters. For smaller discharges, the NRA will ensure that all receive at least fine screening. Whatever level of conventional treatment is applied, further reductions in bacterial concentrations are required to achieve the standards laid down for bathing waters. Bacterial levels in crude sewage are generally around 10,000 times higher than the EC mandatory standards for bathing water quality. Preliminary and primary treatment, whilst achieving major aesthetic improvement, achieve very little in terms of bacterial reduction. Secondary treatment may achieve a one or two order reduction, depending on the particular process, i.e. about the same as the reduction achieved through initial dilution above a modern sea outfall. It is a pre-requisite of modern outfalls that the initial dilution is adequate to avoid the formation of a surface slick. This is achieved by siting the discharge in deep, high energy waters and discharging from a series of ports along a diffuser section. Once the fully mixed buoyant sewage plume has reached the surface, it is driven by the effects of wind and current which disperse the effluent. Enteric micro organisms contained within the sewage, die off in the sea at a rate that is dependent on factors such as temperature and the intensity of incident UV radiation. In consenting a discharge at sea, the NRA uses its knowledge of oceanography and the local wind conditions to ensure that, by the time treated sewage reaches the shoreline, the combined effects of dilution, dispersion and die off will, even under adverse conditions, achieve the necessary reductions in coliform concentrations. It is not possible to consider the main continuous discharge in isolation from the associated overflows. Sewers in the UK are generally combined systems in that they carry both rainwater and sewage. Consequently, after a period of rain, flows of foul water can be very high. Stormwater overflows from the sewerage systems can be highly polluting. The Directive allows results caused by 'exceptional weather conditions' to be disregarded. Her Majesty's Inspectorate of Pollution (HMIP) took the view that this equated to a 1 in 5 year storm event and required sewerage systems to accommodate storms up to this magnitude or to discharge stormwater through overflows well remote from high amenity areas. This policy is under review by the NRA.

Taken together, the conditions required for improved discharges, impose a very substantial cost on the discharger. Most of our sewers and outfalls date back to Victorian times and improving them to accommodate both the sewage flows and the standards of the 1990s, is an expensive task. During the 1990's water companies will spend about £2 billion on improving bathing water quality, with most of the expenditure being during the first five years. The NRA will use the best scientific evidence available to licence these improvements and thereby ensure that the investment is properly targeted in terms of environmental benefit. Applications for consent to discharge will be advertised and account taken of all objections. Once an application has been determined, the discharger has the right of appeal and the Secretary of State has the power to call a public inquiry if there is substantial public opposition to an improvement scheme. All EC bathing waters in England and Wales will comply with the mandatory standards of the Directive by the latter half of the 1990s and, in issuing consents for these improvements, the NRA will ensure that other environmental interests such as marine life and shell fisheries are adequately protected.

REFERENCES

NRA (1991a) National Rivers Authority. *Bathing Water Quality in England and Wales - 1990.* Report of the National Rivers Authority. Water Quality Series No. 3. Published by the National Rivers Authority. 188p.

NRA (1991b) National Rivers Authority. *Proposals for statutory water quality objectives.* Report of the National Rivers Authority. Water Quality Series No. 5 Published by the National Rivers Authority. 188p.

Chapter 4

BATHING WATER QUALITY : A LOCAL AUTHORITY PERSPECTIVE

Steve Oldridge
Environmental Health Department, Great Yarmouth Borough Council

INTRODUCTION

The public health functions of local government were established during the 19th Century as a result of growing awareness that action was required to deal with problems resulting from the rapid growth of industrial and urban society. Local Government at District Council level maintains this strong public health tradition to which has been added responsibility for a wide range of Environmental Protection functions.

Much of the United Kingdom's sewerage and sewage disposal system dates back to Victorian times. In coastal areas topography and cost considerations favoured the discharge of sewage to the littoral zone by means of short outfalls. Underlying this approach was the assumption that the sea killed off bacteria and would sufficiently dilute and dispose of the sewage to render it harmless (HMSO, 1990a). It is only in the latter half of the twentieth century that this assumption has been seriously challenged.

The extent of the problems stemming from this Victorian heritage were reviewed by Kay (1988) who reported that Britain discharged the waste from approximately six million people directly into the nearshore zone. The loading increases during the summer bathing season when daily flow may be in the region of 2.15×10^9 litres of sewage effluent. The potential of this volume of effluent for beach fouling was emphasised by a Department of the Environment report (DoE) which indicated that the majority of British outfalls discharge above low water mark (DoE, 1973).

Problems caused by the discharge of inadequately treated sewage are not, however, confined to Britain and the EC. A United Nations Environment Programme

Report (GESAMP, 1990) confirms that the principal problem for human health, on a world wide scale, is the discharge of pathogens with domestic sewage either directly to coastal waters or to estuaries, rivers and canals that eventually carry such organisms to the sea.

TOURISM CONSIDERATIONS

The wide-ranging responsibilities of Local Government, at District Council level, include not only the protection of public health but also the promotion of tourism, the maintenance of local amenity and environmental protection in a wider sense.

As an industry, tourism has increased rapidly to the point where it is of major significance to the national economy. This growth is reflected in the importance of tourism to the economic well being of seaside towns in Britain. During 1985, sixty six million tourist nights were spent at the top 17 coastal resorts (British Tourist Authority, 1986). The spending associated with this tourism was estimated at approximately £1.7 billion. Scarborough accounted for four million of these tourist nights and it has been estimated that up to 35% of the Borough's employment is directly dependent on tourism with a further 24% being tourist-related to some extent (Ayrton and Oldridge, 1988).

Long before the current widespread public interest and concern about environmental matters, resort towns were aware of the importance of environmental factors to their marketing strategies. For example an English Tourist Board study conducted in May 1981 for Scarborough Council found that a neat and tidy environment was one of the most frequently cited requirements of the tourists who were interviewed (Ayrton and Oldridge, 1988). In 1987 the British Tourist Authority reported that in public opinion surveys about desirable seaside resort characteristics it was usual to find some 10% of respondents citing the importance of a clean beach.

The preamble to the EC Bathing Water Directive acknowledged that interest in the environment and improvement in its quality was increasing (ADC, 1990). Evidence presented in 1985 to the House of Commons Welsh Affairs Committee predicted that the public would become more and more aware of pollution problems and would give greater credence to those authorities which were able to advertise that they had Eurobeaches (HMSO, 1985). Events since then have more than vindicated this prediction.

A DoE survey carried out in 1989 revealed the growth of public concern about the environment. The survey found that sewage contamination of beaches and bathing water was the second most important environmental issue after chemical discharges to rivers and the sea. Approximately 90% of those studied expressed some concern about sewage contamination and 59% reported that they were very worried about it (HMSO, 1990b).

There are other indicators of concern. For example, sales of The Good Beach Guide, published by the Marine Conservation Society, have increased markedly to more than 25,000 copies during the current year (Marine Conservation Society, per comm, 1990).

Interest in the European Blue Flag Award has developed steadily since it was launched in 1987. During 1990 twenty-nine United Kingdom beaches were awarded Blue Flags from a record entry of 63 (EEC, 1990). This year also saw the introduction of the requirement for Blue Flag beaches to show updated information on water quality. The Government has expressed the view that this practice should spread to all Britain's identified beaches (ADC, 1990).

This presents a dilemma for local authorities. A tourism promotion strategy will quite naturally 'accentuate the positive'. This is already seen in the increasing promotion of Blue Flag beaches. At the same time, the potential for bad publicity resulting from publication of water quality results is also recognised. The dilemma is emphasised by the knowledge that widespread publicity about poor water quality may provide the spur to secure improvements to sewage disposal facilities. There is, therefore, a fine balance between the need to avoid bad publicity and the potential to secure early improvement to sewage disposal facilities. It is to be expected, therefore, that authorities, whatever their public stance, will be ambivalent about the recent proposals.

HEALTH RISKS FROM SEWAGE CONTAMINATED BATHING WATERS

The level of health hazard which may be associated with bathing in sewage-polluted water has received a great deal of attention in recent years. The potential relationship between bathing water quality and illness is the aspect of coastal sewage disposal which is of greatest concern to Environmental Health Officers.

It has been suggested that the EC Bathing Water Directive (EEC, 1976) is not primarily concerned with public health but with encouraging long-term improvements in amenity and reassuring the public (Pike and Cooper, 1987). While it is certainly true that the water quality standards specified in the Directive are not based on epidemiological research, such a narrow interpretation would, nevertheless, seem to have little foundation. Although it must be admitted that precise legal interpretation is the function of a Court of Law, the preamble to Directive 76/160/EEC appears to give equal weight to protecting both the environment and public health.

Public concern about poor bathing water quality and failure to meet the Directive standards certainly focuses on worries about health risks. Environmental pressure groups appear to have a similar perspective (The Green Party, 1990). Whatever the legislators' intention, the continuing debate about the most appropriate microbiological monitoring regime and the potential for relating water quality to bather morbidity certainly suggests that the public health significance of the Directive is widely recognised (HMSO, 1990a, Wheeler, 1990, MRC, 1959). Until recently, there have been no epidemiological studies in the United Kingdom concerning the potential health effects of bathing in sewage-contaminated waters. The foundation for policy making has been the acceptance that, except where conditions are so foul as to be aesthetically revolting, the risk to public health from bathing in sea water contaminated by sewage could be ignored. This concept was based on the findings of a retrospective epidemiological study carried out during the late 1950's by the Public Health Laboratory Service (MRC, 1959).

The relevance of the PHLS study has, in recent years, been increasingly called into question. Perhaps the most significant challenge came from the 10th Report of the Royal Commission on Environmental Pollution (RCEP). The RCEP report published in 1984 broadly supported the conclusions of the PHLS study with respect to serious disease, but added the important qualification that

> *there is less confidence that the same can be said of relatively minor infections, ear nose and throat infections and skin and eye irritation (HMSO, 1984).*

The challenge to the PHLS study has, in the main, been based on a series of studies carried out in the USA. Work by Victor J. Cabelli and colleagues at the United States Environmental Protection Agency suggests that there is a risk of illness associated with bathing in sewage-polluted water and the risk increases as water quality decreases (Cabelli, 1983).

Despite the fact that the methodology of the Cabelli studies and their relevance to the colder waters surrounding the United Kingdom has been called into question, a momentum developed which eventually led to the widespread circulation of proposals for a prospective epidemiological pilot survey in the United Kingdom (Jones and Kay, 1988).

In 1987 the Robens Institute undertook the first significant epidemiological study in UK waters. The study demonstrated elevated reporting of gastrointestinal symptoms in bathers compared with non-bathers (Brown *et al.*, 1987).

Work to examine further the validity of Cabelli's findings was subsequently commissioned by the DoE. A pilot study to test a methodology for assessing the relationship between risks to health and bathing was carried out during the Summer of 1989. The findings were published in 1 990 and concluded that the methodology adopted was appropriate but that greatly extended studies would be required to obtain statistically-significant results (WRc, 1990). The House of Commons Environment Committee recommended that these extended studies be undertaken as soon as possible (HMSO, 1990a).

WATER SPORTS

A factor which must be taken into account in any such research programme is the likelihood of individuals who engage in water sports coming into contact with higher levels of pollutants than would be encountered at the shoreline. The increased participation in windsurfing, skiing and jet skiing has caused Scarborough Borough Council to examine the need to designate areas within the inshore zone for such activities. Other local authorities will, no doubt, be in a similar position.

The main purpose for such an exercise is to achieve separation between bathers and water sport activities for safety reasons. This process tends to result in water sports being moved away from the main bathing areas. There is, therefore, a possibility that designated water sport areas will be nearer to outfall mixing zones. Furthermore, water sport activities extend further offshore than do most bathing zones. The water quality sampling regime should have regard to this. The Langland Bay study went part way to addressing the problem in that water quality was assessed at chest height as well as the normal 30cm requirement of the Directive (Jones et al., 1989). However, given that water sport participants may ingest greater quantities of water and come into more violent contact with it than do most bathers, any further epidemiological studies should consider more closely the quality of water specifically designated for such activities.

INLAND WATER

Similar considerations apply to inland waters. Although the EC designated bathing water list for the United Kingdom does not include any fresh water zones, there is no doubt that some inland waters used for recreation would fall within the *bathing water* definition.

The Chartered Institute of Public Finance and Accountancy records that demand for water sports and activities, including water contact sports, has been created almost exclusively in the post war period. The reasons for this are growing pressure from recreational interests, changes in technology, changes in mobility/access and an increase in the ability to pay (CIPFA, 1989). As a consequence of this growth the variety of water sports activities is as diverse as the variety of water on which they are practised. Inland water ranging from rivers and canals to reservoirs and gravel pits are used for recreational purposes. The growth of interest in water recreation activities in both coastal and inland waters shows no signs of diminishing (YHCSR, 1977).

A working group of Environmental Health Officers from the North of England is currently examining the application of microbiological standards to inland waters used for recreational purposes. The group is acting under the auspices of the local authority-based Yorkshire and Humberside Pollution Advisory Council.

Early concern was aroused by a report that a canoe club regularly practised in a river immediately below a major sewage works discharge. Concern was further heightened by the realisation that such practice includes the safety procedure known as the *Eskimo roll*. This example serves to emphasise that the quality of inland water used for bathing and water sport activities must receive greater attention than has hitherto been the case.

PUBLIC INFORMATION ABOUT BATHING WATER QUALITY

An anecdotal comment in a beach quality report published in 1990 by the Green Party (Green Party, 1990) confirms a previously expressed opinion that holidaymakers do not avoid bathing water and beaches which are obviously polluted (Ayrton and Oldridge, 1988). An explanation for this may be that holidaymakers are able to recognise pollution but are not sufficiently concerned to avoid such areas. This would seem to be most unlikely. It is probably the case that most simply do not recognise obviously polluted situations and are, therefore, unable to exercise choice. The publication of bathing water quality results would, therefore, provide this freedom of choice.

However, it is important that people be able to make an informed choice. Simply presenting monitoring data on total coliform and faecal coliform counts for the current year or previous years will not, by itself, achieve this. It is, therefore, necessary that the data be provided with a degree of explanation which is both consistent and accurate. The format currently used for Blue Flag beaches attempts this by summarising water quality for each of the previous four years as either Excellent, Good, or Unsatisfactory. While these descriptions are consistent, it may be that wider application of the same format would create difficulties because the Unsatisfactory classification would be applied to beaches which range from marginal failure to grossly polluted, i.e. everything between 0% and 90% compliance.

The Association of District Councils is currently discussing with the National Rivers Authority how best the requirement to publicise water quality results can be implemented (ADC, 1990). It is, therefore, important that the question of interpretation be addressed.

Furthermore it is likely that when information about water quality becomes more widely available demand for beach closure during periods when water quality is 'unsatisfactory' will increase. Kay (1988) has pointed out that the present sampling regimes in Europe do not provide the management information on which to base such a policy. It is, therefore, essential that effort be directed towards establishing both relevant faecal indicator systems and the relationship, if any, between water quality and bather morbidity. Such studies will have very significant implications for both the public health and tourism concerns of coastal local authorities.

Over the years a number of different approaches to pollution control have been evolved by national governments and international organisations. Such strategies do not, however, evolve in a vacuum. In common with all 'social policy' initiatives they are influenced by social, political and economic factors which, together, help shape the national philosophy and approach to matters such as pollution control. While environmental policy must be based on good science it is important to recognise that science is not the final arbiter of such policy.

In view of recent changes in United Kingdom policy towards discharges to the marine environment and pending EC changes in the form of the proposed Municipal Waste Water Directive, it is worth examining some of the concepts which may underlie pollution control strategy. Although there are problems of definition, the following three approaches can be identified:

(i) the anticipation principle,

(ii) emission standards,

(iii) water quality objectives and standards.

THE ANTICIPATION PRINCIPLE

The anticipation principle is founded on the idea that prevention is better than cure and, accordingly, emissions should be prevented without waiting for evidence of damage. The EC's first action programme on the environment included this concept (EEC, 1973:112):

> *The best environmental policy consists in preventing the creation of pollution or nuisances at source, rather than trying to counteract their effects.*

However, the EC document went on to qualify this statement by a reference to such policy being compatible with economic and social development.

A more stringent version of the anticipation principle is embodied in the *Vcrsorgeprinzip* of the Federal Republic of Germany. This principle was enunciated in the context of the debate about pollution of the North Sea and accords environmental protection higher priority than other costs in society (Weizsaker *et al.*, 1988).

While there is no definitive statement of the anticipation principle, it does embody the concepts embraced by environmentalists in many countries (Peet, 1988). For scientists and policy makers it poses the question; does what we now know about the environment form an adequate basis for action or should it be assumed that the future holds, as yet, unidentified risks which current policy should address. By implication the anticipation principle poses, but leaves unanswered, the question of what degree of cost society is prepared to pay to avoid risk.

EMISSION STANDARDS

A widely used approach to controlling the emission of pollutants from a point source is to impose emission standards. Standards may be established by reference to the quality of the receiving environment, to what is achievable given current or reasonably foreseeable technology, or a combination of these factors.

Emission standards may be uniform and applicable to all emissions of a similar character, or non-uniform. An advantage of applying a fixed standard to all similar emissions is that all manufacturers are treated equally whereas non-uniform standards may take account of the vulnerability of particular environmental circumstances.

An emission standard approach does, however, have several drawbacks. It cannot deal with non-point source emissions such as agricultural run-off or atmospheric deposition to water (both of which are important factors in a North Sea context): it cannot deal with the effect of too many point sources which may, in combination, overload the environment, and it requires a procedure to be established for updating standards in line with technological improvements (Weizsaker *et al.*, 1988).

From the environmentalist's viewpoint reliance on emission standards as a method of pollution control may be criticised on the basis that such an approach will do no more than is strictly necessary to prevent unacceptable interference with desired uses of the media to be protected.

ENVIRONMENTAL QUALITY OBJECTIVES

A pollution control methodology which takes into account non-point source emissions is the use of Environmental Quality Objectives (EQO) and the imposition of related Environmental Quality Standards (EQS). Such an approach can have regard to the fact that it may be acceptable for water quality levels to vary from place to place depending on the different uses of the water.

A disadvantage of using EQOs/EQSs is that they will provide a meaningful guide for fixing individual emission standards from point sources only where reliable cause and effect relationships can be established. The larger the body of water the less reliable are EQOs/EQSs as a guide to establishing consent standards for an individual source. Nevertheless, advocates of the United Kingdom approach to water pollution control may point to improvements in the condition of rivers and estuaries as evidence that the use of EQOs/EQSs together with non-uniform emission standards, are a cost-effective method of securing improved environmental standards (Downing, 1988).

NEW CHALLENGES

The EC Bathing Water Directive (76/160/EEC) is an example of a control measure which uses specific environmental quality standards to achieve an environmental quality objective, namely the fitness of recreational waters for bathing.

Whilst the relevance of the current Bathing Water Directive standards to the protection of public health and, indeed, the purpose of the Directive, have been debated, it does at least provide a fixed objective to attain bathing water of a certain minimum quality.

THE PROPOSED MUNICIPAL WASTE WATER DIRECTIVE

In addition to the requirements of the Bathing Water Directive it is now clear that the proposed EC Directive concerning municipal waste water treatment (EEC, 1990) will specify the standard of treatment of sewage and similar industrial waste discharges. The Waste Water Directive can, therefore, be seen as a move away from the EQO/EQS philosophy of the Bathing Water Directive in favour of an Emission Standard approach.

The proposed Directive is not yet in its final form (December 1990) but it is clear that at least primary and secondary treatment will be required for virtually all sewerage discharges from populations greater than 2,000. In addition, the 'sensitive area provisions' proposed in Article 5 will require nutrient removal from many such discharges.

Article 6 of the proposal permits member states to identify 'less sensitive' areas. Sewage discharges to such areas may be subjected to less stringent treatment than that referred to above provided that they receive at least primary treatment, and comprehensive studies indicate that such discharges will not adversely affect the environment. It is likely that the environmental lobby, which tends to favour the precautionary approach, would resist large-scale designation of 'less sensitive' areas.

The United Kingdom Government has already accepted, in principle, the requirements of the EC Draft Municipal Waste Water Directive. In fact, the commitment of the Government is that in future all significant discharges of sewage should be treated at a sewage treatment works before discharge to estuarine or coastal waters (DoE, 1990). To achieve this a £2.9 billion investment programme has been announced. Given this commitment, which was confirmed in the recent White Paper on the Environment (HMSO, 1990c), and the strength of the environmental lobby, it seems unlikely that the Government has much scope to make widespread use of the 'less sensitive' area provisions contained in the Draft Directive.

A NEW POLICY

Prior to the Government's acceptance of treatment for coastal and estuarine sewage discharges, United Kingdom policy was consistent with the findings of the Jeger report that

> *sewage should only be discharged to sea after screening, comminution and through diffusers on long outfalls (HMSO, 1970).*

This finding was endorsed by the 1984 Royal Commission on Environmental Pollution which concluded that

> *with well designed sewage outfalls ... the discharge of sewage to the sea is not only acceptable but, in many cases, environmentally preferable to alternative methods of disposal (HMSO, 1984).*

The Royal Commission criticised the Government for the slow implementation of the Jeger report. In their response, the Government agreed that the elimination of crude sewage discharges was a long-term aim but, at the same time, was reluctant to set a target date for elimination of all such discharges (DoE, 1984).

It must remain a matter of conjecture whether the pollution control philosophy embodied in the proposed Municipal Waste Water Directive was influenced more by the increasing importance of environmental matters to the political process than by scientific and technical considerations. The proposal does, nevertheless, render untenable the policy of dealing with outfalls on an individual basis taking account of Water Quality Objectives and the characteristics of the receiving water.

Against this background it is, therefore, not surprising that environmental groups have hailed the Government's recent announcements about the treatment of coastal sewage discharges as a victory. Many local authorities have been at the forefront of the campaign to secure improved sewage disposal arrangements and have also warmly welcomed the proposals. At the same time, however, there is a growing realisation that the change in policy will present new challenges to local authorities.

These relate to the siting and operation of sewage works together with the associated problem of sludge disposal.

PUBLIC PERCEPTIONS

It is perhaps not too much of an exaggeration to claim that the general public perceive a sewage works as a system where dirty water goes in and clean, if not quite drinkable, water comes out. The problems of sludge disposal and the need for a suitable watercourse for dilution of the final effluent are not widely recognised: nor is the physical disruption and other problems associated with reversing the direction of flow of Victorian sewerage systems to cause them to discharge either by gravity or via pumped systems, away from the coast. Construction of an inland sewage works may, therefore, be seized upon as a complete and uncomplicated answer to unsatisfactory sea outfalls.

The reality is somewhat different. Secondary treatment, preceded by primary treatment, reduces faecal coliforms to between 10% and 50% of their original concentration in raw sewage and viruses to between 10% and 40% of their original concentration. The reduction in micro organisms is a function of the treatment method used. Typically, therefore, a 90% reduction of faecal coliforms after secondary treatment still leaves a concentration of about one million faecal coliforms per 100 ml of treated effluent. A further 500-fold dilution would, therefore, be required to reach the Bathing Water Directive standard assuming no die-off with time. On the basis that an initial dilution of 1:10 would be typical of a short outfall in fairly turbulent water, the Water Research Centre has pointed out that an outfall of a substantial length would still be required in order to meet the EC standard (WRc, 1977).

Whilst disinfection may be proposed as a means of reducing the bacterial loading after secondary treatment, there are drawbacks associated with it (Pike and Cooper, 1987). The use of oxidising agents, particularly chlorine, gives rise to concern about the formation of potentially harmful organic compounds. Accurate dosing is a problem and this is emphasised during high flows and storm conditions. Techniques such as ozonation and ultra-violet treatment have cost and technical problems. Other processes are available but, as yet, disinfection is not practiced to any significant effect in the country.

It must be recognised, however, that disinfection may have a greater part to play in the future. Furthermore it has been pointed out that disinfection techniques may also have beneficial effects on water quality both in respect of shell fisheries and water contact sports (HMSO, 1990a).

Nevertheless, whatever is done by way of treatment, topography will determine that in the vast majority of coastal situations final effluent will have to be discharged directly to the sea. Careful attention will, therefore, still have to be given to the design and siting of sea outfalls and diffuser systems to replace existing short outfalls which discharge in or near to the inter-tidal zone.

THE NIMBY SYNDROME

The problem of disposing of sewage sludge from inland sewage works is already causing water companies to plan disposal strategies which are less dependent on the use of agricultural land (Hudson and Walker, 1990). The creation of additional sludge by coastal sewage treatment and the cessation of sewage sludge disposal at sea by 1998 will exacerbate these difficulties.

In the context of the predicted need for additional sludge incineration capacity, the Environment Committee Report pointed out the problems experienced in the past in securing planning permission for sludge incinerators (HMSO, 1990a). The report also quoted the DoE view that sewage treatment facilities constructed as an alternative to long sea outfalls would inevitably involve problems related to odour, sludge disposal and nuisance in their own right.

The challenge the new policy will present to local authorities was recently emphasised by reports that the Environment Minister, David Trippier, was concerned that the Not In My Back Yard (NIMBY) principle could stop local authorities giving planning permission for new sewage treatment plants (Mail on Sunday, 1990).

The House of Commons Environment Committee report contains a reference to public expectations about alternatives to marine disposal of sewage:

> *It was equally clear that there was a widespread illusion that other methods of sewage treatment could be implemented without any cost to the environment and utilisation of land and other resources (HMSO, 1990a).*

The Government's recent White Paper on the environment emphasised that the answers to environmental problems are not straightforward. It asserted that the way to earn public confidence in the approach to environmental policies is to tell people the facts and what they mean and to give them the opportunity to make their views known (HMSO, 1990c).

It is clear that in environmental terms the alternatives to the marine disposal of sewage are not straight forward. The massive capital investment programme in land-based treatment facilities will bring its own problems. The NIMBY syndrome will inevitably emerge. Nevertheless the programme to eliminate unsatisfactory discharges must proceed as speedily as possible. The planning process will ultimately decide the fate of particular schemes. However, Environmental Health Officers will have an important part to play in ensuring that the issues raised are debated in an informed way so that conflicting environmental interests can be resolved.

CONCLUSIONS

The issue of bathing water quality continues to be a matter of concern to local authorities from both the public health and tourism perspectives. Recent changes in Government policy regarding the disposal of sewage to the marine environment and the proposed EC Municipal Waste Water Directive will present new challenges.

The proposal to publish information about bathing water quality will further increase the level of public awareness and concern. Questions about the safety of bathing in sewage-polluted waters will be raised and demands for closure of 'unsatisfactory' beaches will increase. The relevance of the parameters monitored to assess compliance with the Bathing Water Directive is increasingly being called into question. Furthermore the current state of epidemiological knowledge is inadequate to provide meaningful answers to questions about the public health implications of sewage-contaminated bathing waters.

There is, therefore, an urgent need for the completion of epidemiological studies. Such studies should examine potential risks to both bathers and those who engage in water sports. The need to relate such studies to recreational activities in inland waters should also be addressed.

The proposals for an EC Municipal Waste Water Directive and recent Government announcements about the discharge of untreated sewage to the marine environment mean that at least primary and secondary treatment of all significant discharges of sewage from coastal communities will become the norm. This policy will not, however, obviate the need to deal with existing short outfalls which discharge in the inter-tidal zone.

Both the construction and operation of new coastal sewage treatment systems will give rise to environmental problems. It is likely that sludge disposal will be a particular source of concern.

The NIMBY syndrome may result in excessive delays at the planning permission stage of development. It is, therefore, essential that an informed and widespread debate be initiated aimed at reconciling the need for improved coastal sewage disposal facilities with the consequences of such a policy. Local government and, in particular, Environmental Health Officers, will have an important role to play in addressing the conflicting environmental issues which such a debate will raise.

REFERENCES

ADC (1990) Association of District Councils. Circular 1990/192 25th July.

Ayrton, W.R., Oldridge, S. (1988) *Littoral zones, amenities and tourism in environmental protection of the north sea.* in Newman, P.J. and Agg, A.R. (Eds) *Environmental Protection of the North Sea.* Heinemann.

British Tourist Authority (1986) *British Tourist Authority Statistics 1985.* Englands Domestic Seaside Tourism and Top English Towns for Domestic Tourism 1985-86.

Brown, J.M., Campbell, E.A., Rickards, A.D. and Wheeler, D. (1987) *The Public Health Implications of Sewage Pollution of Bathing Water.* The Robens Institute of Industrial and Environmental Health and Safety. University of Surrey.

Cabelli, V.J. (1983) *Health Effects Criteria for Marine Recreational Water.* Report EPA 600/1-80-031. US Environmental Protection Agency, Ohio.

CIPFA (1990) Chartered Institute of Public Finance and Accountancy (CIPFA). *Financial Information Service 19*, CIPFA 1989.

DoE (1984) Department of the Environment. *Pollution Paper No. 22.* HMSO, London.

DoE (1990) Department of the Environment. *News Release* 5th March 1990.
Downing, A. L. (1988) *Quality Objectives and Discharge Consents*. Chapter 31 in in Newman, P.J. and Agg, A.R. (Eds) *Environmental Protection of the North Sea. Heinemann.*

EEC (1973) *Official Journal of the European Communities. 20.12.73.* C.112.

EEC (1990) *Official Journal of the European Communities. No. C.1 4th January 1990.*

EEC (1990) *The European Blue Flag 1990.* Commission of the European Communities, Brussels.

GESAMP (1990) Joint Group of Experts on the Scientific Aspects of Marine Pollution. *The State of the Marine Environment.* UNEP Regional Seas Reports and Studies No. 115. UNEP 1990, paragraph 217.

Green Party (1990) *The State of Britain's Beaches 1990.* The Green Party, London 1990.

HMSO (1970) *Taken for Granted: Report of the Working Party on Sewage Disposal.* HMSO London.

HMSO (1973) *Report of a Survey of the Discharges of Foul Sewage to the Coastal Waters of England and Wales.* DoE. HMSO London.

HMSO (1984) Cmnd.9149. Royal Commission on Environmental Pollution *Tenth Report.* HMSO London.

HMSO (1985) House of Commons Committee on Welsh Affairs. *Coastal Sewage pollution in Wales*: Minutes of Evidence. HMSO London.

HMSO (1990a) House of Commons. Session 1989-90, *Fourth Report - Pollution of Beaches.* HMSO London.

HMSO (1990b) Department of the Environment. *Digest of Environmental Protection and Water Statistics No. 12.* HMSO London.

HMSO (1990c) Cm 1200. *This Common Inheritance - Britain's Environmental Strategy.* HMS0 London.

Hudson, J.A. and Walker, J.B. (1990) *Sewage Sludge Incineration a Strategy for the Future.* A paper presented to an open meeting of the North East Centre of the Institute of Wastes Management. Bradford 27th April.

Jones, F. and Kay, D. (1988) *Joint Research Proposal on Coastal Bathing Related Disease Incidence.* Altwell Hygiene and Environmental Consultant Analysts, Warrington and St Davids University College, Lampeter, 1988.

Jones, F., Kay, D., Stanwell-Smith, R. and Wyer, M.D. (1989) *The Langland Bay Controlled Cohort Pilot Study.* Centre for Research into Environment and Health, University of Wales. Lampeter.

Kay, D. (1988) Coastal Bathing Water Quality: The Application of Water Quality Standards to Welsh Beaches. *Applied Geography 8,* 117-134.

Mail on Sunday (1990). 18th November 1990.

Marine Conservation Society (1990). Personal communication 18th November.

MRC (1959) Medical Research Council Memorandum No. 37. *Sewage Contamination of Bathing Beaches in England and Wales.* HMSO London.

Peet, J.G. *The Anticipation Principle as a Basis for Policy.* Chapter 30 in in Newman, P.J. and Agg, A.R. (Eds) *Environmental Protection of the North Sea.* Heinemann.

Pike, E.B. and Cooper, V.A. (1987) *Control of Pollution in Recreational Waters - The Way Forward.* Paper presented to the Institution of Environmental Health Officers at a seminar *'Preventing Water Pollution - The Environmental Health Aspect'* - Blackpool Branch. Mimeographea.

Weizsacker, E., Moltke, K., and Haigh, N. *Towards an Integrated Approach.* in Newman, P.J. and Agg, A.R. (Eds) *Environmental Protection of the North Sea.* Heinemann.

Wheeler, D. (1990) The Risk of Bathing in Water Contaminated by sewage. *Environmental Health 98(10),* 285-287.

WRc (1977) Water Research Centre. *Design Guide for Marine Treatment Schemes:* Volume 1. WRc Medmenham.

WRc (1990) Water Research Centre. *Health Effects of Sea Bathing Phase 1*. WRc 1990.

YHCSR (1977) Yorkshire and Humberside Council for Sport and Recreation. *Strategy for Water Recreation in North Yorkshire*. YHCSR Leeds 1977.

Chapter 5

LEGAL ISSUES CONCERNING BATHING WATERS

William Howarth
Director of the Centre for Law in Rural Areas and Senior Lecturer in Law, University College of Wales, Aberystwyth.

INTRODUCTION

The European Community's Bathing Water Directive (EEC, 1976) establishes standards for the quality of bathing water and gives rise to corresponding duties upon the twelve Member States of the Community who are obliged to achieve and maintain compliance with the objectives of the Directive through the enactment of national laws and the establishment of appropriate administrative structures. This Chapter seeks to present the key legal features of the Bathing Water Directive in its European context, and to give particular consideration to those legal and administrative measures in England and Wales which seek to ensure conformity with its requirements.

THE EUROPEAN DIMENSION TO BATHING WATER QUALITY

Initially, the legal status of the Bathing Water Directive as a *Directive* is to be noted. The significance of this form of legislation is that it is

> *binding as to the result to be achieved . . . but shall leave to the national authorities the choice of form and methods (EEC, 1976).*

This means that although a Member State is bound to realise the objectives set out in a Directive, it is allowed some flexibility in the manner of implementation to take account of different national administrative and legal circumstances. However, the overriding obligation upon the Member State remains that of securing that a positive legislative act of implementation by the appropriate national authority takes place within the period stipulated by the Directive unless, of course, the objectives of the Directive already have force of law in the Member State concerned. Hence, in the United Kingdom, s.2(3) of the European Communities Act (1972) allows for the implementation of Directives by means of delegated legislation. Accordingly, it would be open to the appropriate Minister to introduce a statutory Regulation or Order to implement a Community Directive if the legal means for implementation of that Directive is not otherwise available in United Kingdom law.

A notable contrast between Community and domestic United Kingdom Law concerns the consequences of breaches of the different levels of law. Clearly breach of the environmental laws of the United Kingdom will make the transgressor liable to whatever punishments the law provides for. In contrast, however, Community Directives rarely impose obligations directly upon individuals. Because of this, legal proceedings for failure to implement a Directive are brought by the Commission against the government of the Member State concerned. The European Commission has both the power and duty to enforce Community legislation, and where the Commission considers that a Member State has failed to fulfil a Community obligation, such as that of implementing a Directive, the Commission is bound to deliver a reasoned opinion to the State concerned giving it the opportunity to submit its observations. If the State does not comply with the opinion of the Commission within the period laid down by the Commission the latter may bring the matter before the European Court of Justice.[1] Alternatively a Member State which considers that another Member State has failed to fulfil an obligation may bring the matter before the Court of Justice, but the matter must first be brought before the Commission and the Commission is bound to deliver a reasoned opinion after each of the States concerned has been given the opportunity to submit its own case and its observations on the other party's case.[2]

By whichever route a complaint passes to the European Court, if the Court finds that a Member State has failed to fulfil an obligation required by the Treaty that State is required to take the necessary measures to comply with the judgement of the Court.[3] Significantly, however, the Treaty does not provided for any coercive means by which adherence to the obligation can be secured. The judgements of the Court are, therefore, declaratory rather than mandatory in legal character. It must be added, however, that disobedience to a ruling of the Court would carry strong political pressure for conformity with Community Law, and sustained intentional disobedience to the Law might in the last resort be seen as inconsistent with continuing membership of the Community. These matters are, however, theoretical rather than practical considerations since Member States are keen to avoid being brought before the Court to answer charges concerning their environmental records and seek to comply with resolutions of the Court. Hence a member State will normally acquiesce to the resolutions of the Court to avoid damage either to its national environmental

---

[1] *Ibid.* Art..69.
[2] *Ibid* Art 170.
[3] *Ibid* Art 171.

reputation or to its apparent commitment to Community environmental objectives.

Despite the general ethic of conformity to Community Law for political reasons, transgressions of Environmental Directives, including the Bathing Water Directive, have been brought before the European Court in the past. In *Commission of the European Communities v. Kingdom of the Netherlands*[4] the Court found that though each Member State is free to implement a Directive by means of measures adopted by regional or local authorities, the overriding obligation remained that of giving effect to the Directive by means of legally binding provisions. On the facts of the case, the mere changes of administrative practice which had been adopted by the Dutch Government did not constitute implementation of the Directive and the Court found them guilty of failing to implement it. More recently infringement proceedings have been pursued by the European Commission against all the Member States of the Community in relation to bathing waters, except Portugal [5] which has been given until 1 January 1993 to comply.[6] Infraction procedures are under way against the United Kingdom in respect of the beaches at Blackpool, Formby and Southport where coliform counts, indicating sewage contamination, have failed to reach standards required by the Directive.[7] It would seem, therefore, that general difficulties have been encountered in meeting the requirements of the Directive throughout the Community.

BATHING WATERS UNDER THE BATHING WATER DIRECTIVE

The detailed Community obligations imposed upon the United Kingdom Government in relation to bating water are set out in the Bathing Water Directive. This Directive was enacted under the Treaty powers provided by Articles 100 and 235. These provisions allow respectively for the Community to take action where necessary for the establishment or functioning of the common market, and for action to be taken by the Community where it is necessary to attain, in the course of operation of the common market, an objective of the Community for which the Treaty has not provided the necessary powers.[8] Although doubts have been expressed in the past about the use of legislative powers concerned with the functioning of the common market to achieve environmental objectives,[9] the unanimity with which the Directive

---

[4] Case 96/81 [1982] ECR 1791.
[5] House of Commons Environment Committee, Fourth Report, *Pollution of Beaches*, (1990) House of Commons Paper 12-I.
[6] OJ 1990 C125/13.
[7] [1990] *Water Law* 32.
[8] Contrast the powers to enact environmental legislation now provided for by the Single European Act 1987 which insets new Articles into the 1957 Treaty allowing for environmental legislation: see Articles 130R to 130T of the Treaty as amended.
[9] See House of Lords Select Committee on the European Communities, *Approximation of laws under Article 100 of the EEC Treaty*, 22nd Report Session 1977-78; and House of Lords' Select Committee on the European Communities, *Environmental Problems and the Treaty of Rome*, Session 1979-80.

was enacted by the European Council of Ministers has meant that the constitutionality of this Directive, and other environmental measures, was never seriously challenged. Nonetheless the explicit provision for future environmental legislation now provided for under the Single European Act (1987) is to be welcomed in putting to an end the uncertainties engendered by an economic Community legislating in non-economic areas.[10]

The Bathing Water Directive notes in its preamble that in order to protect the environment and public health it is necessary to reduce the pollution of bathing water and to protect such water against further deterioration. Notably, however, the Directive is selective in the way that it seeks to achieve these objectives in protecting only those parts of the aquatic environment that constitute bathing waters, perhaps because separate Directives are concerned with the protection of other features of the aquatic environment.[11] Significantly the Directive is silent in relation to the preservation or enhancement of the amenity value of bathing waters, and it would seem that the provision of recreational facilities is not an explicit part of the rationale of the Directive. It appears also that it is the *conjunction* of the protection of the environment and public health that is envisaged as the basis of the Directive in that water intended for therapeutic purposes and water used in swimming pools is specifically excluded from the ambit of the Directive.[12]

Unlike other Directives which leave to Member States the discretion to *designate* waters for particular purposes,[13] the Bathing Water Directive is couched in terms of an *objective* definition of the meaning of *bathing water*: waters either are, or are not, bathing waters within the definition of the Directive.[14] Accordingly, Member States have no discretion to redefine *bathing waters* to suit their preference but are bound to apply the provisions of the Directive to those waters which come within the defined meaning of the phrase. Unfortunately, however, the history of the Directive has shown that the definition provided by the Directive is so imprecisely worded that in fact, if not in law, the term is subject to widely differing interpretations.

In the words of the Directive, *bathing water* is defined to mean all running or still fresh waters or parts thereof and sea water, in which: bathing is explicitly authorised by the competent authorities of the member state, or in which bathing is not prohibited and is traditionally practised by a large number of bathers.[15] The wording of this definition given rise to a number of points of legal significance concerning

---

[10] See Vandermeersch, "The Single European Act and the Environmental Policy of the European Economic Community", 12 *European Law Review* 407, at p.411.
[11] See, for example, the Directive on the quality of fresh waters needing protection or improvement in order to support fish life, 78/659/EEC.
[12] Art.1 Bathing Water Directive.
[13] Contrast the power of designation of waters provided for under the Directive on the quality of fresh water needing protection or improvement in order to support fish life, 78/659/EEC; and the Directive on the quality required for shellfish water, 79/923/EEC.
[14] Haigh, *EEC Environmental Policy and Britain* (2nd Ed. 1987) p.63.
[15] Art.2(a) Bathing Water Directive.

the meaning of the terms used. First, it is to be noted that bathing in watercourses, still waters and the sea is envisaged, but in each case at least one of two criteria must be satisfied: *either* that the waters are waters where bathing is explicitly authorised by the competent authority of the Member State *or* waters in which bathing is not prohibited and is traditionally practised by a large number of bathers. Second, at common law the public right to the use of coastal waters extends only to the rights of navigation and fishery. There is no common law *right* to bathe over the foreshore in the coastal waters of the United Kingdom,[16] but equally it would be rare to find the activity prohibited in law. Local authorities are provided with a power to make byelaws with respect to public bathing, to regulate the areas in which, and the hours during which, public bathing is permitted. Conversely, local authorities may prohibit or restrict public bathing at times and places as respects which warning is given by the display of flags, or by other means specified in the byelaws, that bathing is dangerous.[17] Despite these statutory provisions, the view has been taken that the powers of local authorities in this respect limit an existing entitlement rather than creating a new entitlement. For that reason the Government has maintained that there are no bathing waters in the United Kingdom where bathing is explicitly authorised, and most other Member States have taken the same view.[18] Because there are no waters in the UK where bathing is explicitly authorised, reliance is placed exclusively upon the second limb of the Directive definition, so that *bathing waters* is officially understood in the UK to mean sea water where bathing is not prohibited and is traditionally practised by a large number of bathers. Finally, having settled upon the criterion of bathing being practised by a large number of bathers, the outstanding difficulty is that of determining what is to count as *a large number of bathers*.

The imprecise meaning of "bathing waters", and not least the uncertainty as to what was to count as "a large number of bathers", allowed the United Kingdom to adopt an initial approach to implementation of the Directive which showed a rather shameful disregard of the policy objectives of the Directive in the face of economic reservations about implementing it. In 1979 the United Kingdom Government took the view that only coastal waters would fall within the Directive since bathing was neither authorised nor traditionally practiced by large numbers of bathers in freshwater. Thereafter there were relatively few coastal waters where bathing was traditionally practiced by large numbers of persons, as opposed to popular beaches frequented by large numbers of persons *who did not actually venture into the water.*[19]

---

[16] *Brinckman v. Mately* [1904] 2 Ch.313.
[17] s.231(1)(a) and (aa) Public Health Act 1936. The local authority concerned may exercise the powers provided in relation to public bathing in respect of any area of the sea which is outside its area but within 1,000 metres to seaward of any place where the low water mark is within or on the boundary of the authority's area (s.17(1) Local Government (Miscellaneous Provisions) Act 1976).
[18] Haigh, *EEC Environmental Policy and Britain* (1987) p.66.
[19] Department of the Environment, Advice Notices of March 1977 and August 1979, discussed by Haigh, *Ibid*.

As a result of this unduly restrictive approach towards the identification of bathing waters and excessive figure stipulated as the numbers of persons needed to be present to constitute a *large* number of bathers,[20] only 27 beaches were initially listed, excluding the popular bathing beaches at Blackpool and Brighton.

The initial approach of the United Kingdom Government towards the implementation of the Bathing Water Directive was the subject of unreserved criticism by the Royal Commission on Environmental Pollution in their Tenth Report in 1984, who criticised the "unrealistic" criteria for the selection of bathing waters.[21] Eventually the Government accepted this criticism and in 1987 announced the identification of further bathing waters within the terms of the Directive to bring the total up to 391.[22] Presently there are 446 bathing beaches within the terms of the Directive, of which 77% complied with the water quality requirements.[23] Clearly the Directive would have been less controversial if the meaning of *bathing waters* had been more precisely specified from the outset, and the difficulties to which the failure to do so gave rise provide a clear lesson for the drafting of future Directives.

WATER QUALITY AND THE BATHING WATER DIRECTIVE

Having determined which waters are to be subject to the Directive, the substantive obligations imposed by the measure oblige Member States to set quality standards for each individual bathing area in their state with values taken from the parameters provided in the annex to the Directive. Most significant of the quality standards are concerned with the bacteriological content of the water, and in particular with the coliform values indicating the degree of sewage contamination of waters. Quality standards for designated bathing waters must be no less stringent than the *imperative* or *I* values set out in the Annex to the Directive, and member states are also to endeavour to observe the *guide* or *G* values set out in the Annex.[24] Bathing water conforms with relevant parameters where samples of the water meet the parametric values for the water quality for 95 per cent of samples in respect of imperative values, and for 90 per cent of samples in other cases, except for certain coliform parameters in respect of which conformity is set at 80 per cent.[25]

Member States were to have taken all the necessary measures to ensure that, compliance with the Directive was secured by 1985, but derogations from the water quality standards for bathing waters may be permitted where these are based upon plans for the management of the water within the area concerned.[26] The Directive

---

[20] Royal Commission on Environmental Pollution, Tenth Report, *Tackling Pollution - Experience and Prospects*, (1984 Cm.9149) p.89.
[21] *Ibid.* para.4.69.
[22] Department of the Environment, Pollution Paper No.22, *Controlling Pollution: Principles and Prospects* (1984) para.33.
[23] *Water Bulletin*, 23 November 1990 p.5.
[24] Art.3 Bathing Water Directive.
[25] Art.5(1).
[26] Art.4.

may also be waived in respect of certain parameters because of exceptional weather or geographical conditions, or because water undergoes natural enrichment by receipt from the soil of substances contained therein.[27] Alternatively, Member States may at any time fix more stringent values for bathing water than those laid down by the Directive.[28]

Competent authorities in member states are to carry out sampling operations in a prescribed manner at places where the daily average density of bathers is greatest at frequencies determined by the annex to the Directive.[29] In the event of inspection or sampling revealing a discharge of a substance likely to lower the quality of bathing water, or if there is any other grounds for suspecting that there is a decrease in the water quality, additional sampling is also to take place.[30]

Although the Bathing Water Directive is clearly of considerable importance in relation to maintenance of water quality in coastal waters it should not be regarded in isolation from the context of a body of Community legislation concerned with the aquatic environment whereby the common objectives of maintaining and improving the quality of the aquatic environment are sought through a range of different measures. Of particular note in this context are measures such as the Shellfish Waters Directive,[31] which allows for the designation of waters as needing protection or improvement in order to support shellfish and so to contribute to the high quality of edible shellfish products; and the Dangerous Substances Directive,[32] which seeks to establish a framework for the elimination or reduction of pollution by certain specified dangerous substances in inland, coastal and territorial waters. In addition, the Environmental Assessment Directive,[33] which requires certain public or private projects which are likely to have a significant effect on the environment to be subject to environmental assessment, may be particularly relevant to projects such as the construction of new sea outfalls from sewage treatment works. In the future these measures will be significantly augmented by the present Draft Directive on Municipal Waste Water,[34] which envisages minimum requirements for the treatment of municipal waste water and the disposal of sludge to prevent the environment form being adversely affected by municipal waste water discharges. The combination of these and other Directives should be seen as a web of interconnected measures which in differing ways seek to maintain or improve the quality of the aquatic environment.

---

[27] Art.8.
[28] Art.7.
[29] Art.6(1) and (2).
[30] Art.6(4).
[31] Directive on the quality required for shellfish waters, 79/923/EEC.
[32] Directive on pollution caused by certain dangerous substances discharged into the aquatic environment of the Community, 76/464/EEC.
[33] Directive on the assessment of effects of certain public and private projects on the environment, 85/337/EEC.
[34] COM (89) 518 final.

IMPLEMENTATION OF THE DIRECTIVE IN THE UNITED KINGDOM

Having noted the content of the Bathing Water Directive and the obligations that it imposes upon the Government of the United Kingdom the next issue becomes that of how the Directive is implemented within the United Kingdom. The process of implementation involves the identification of a national administrative body as the competent authority for the purpose of ensuring internal compliance with the Directive within the United Kingdom and a body of domestic law which serves to place upon those with an interest in bathing waters obligations which are at least as strict as those arising under the Directive.

In so far as coastal bathing waters are concerned, the Community objective of securing water quality standards is a matter of some administrative complexity when the collage of jurisdictions over coastal waters in the United Kingdom is recognised. Considering the position in England and Wales, and discounting private interests which may arise in relation to the use of coastal waters, the following administrative bodies may be concerned with maintenance of bathing water quality. Local authorities, through their planning departments, have control over developments taking place in the coastal zone as far seaward as the low water mark, and through their environmental health departments, have public health responsibilities which in some respects extend seaward beyond the low water mark.[35] Responsibilities in relation to fisheries are entrusted to Sea Fisheries Committees up to the three nautical mile limit,[36] and for salmon and freshwater fisheries are allocated to the National Rivers Authority up to the 6 mile limit. Beyond those limits, to the extent of the 200 mile Exclusive Economic Zone fisheries responsibilities and powers in relation to the deposit of waste at sea rest with the Ministry of Agriculture, Fisheries and Food.[37] The Nature Conservancy Council has powers in relation to the designation of marine nature reserves in territorial waters up to the 12 mile limit. Seabed ownership up to the 12 mile limit rests with the Crown Estate Commissioners, though the Department of Energy has various responsibilities in relation to offshore oil and gas licensing up to the 200 mile limit. In their different ways each of these concerns has various rights or duties in relation to coastal or sea water quality. Despite the range of diverse bodies with interests in these waters the main body concerned with the maintenance of bathing water quality is the National Rivers Authority.

---

[35] "Coastal waters", for these purposes, means waters within 3 nautical miles from any point on the coast measured from low water mark of ordinary spring tides (s.343(1) Public Health Act 1936).

[36] The Sea Fisheries Regulation Act 1966 gives powers to local sea fisheries committees to make byelaws. Specifically a local fisheries committee may, subject to regulations made by the Minister of Agriculture, Fisheries and Food, make byelaws for the purpose of prohibiting or regulating the deposit or discharge of any solid or liquid substance detrimental to sea fish or sea fishing (s.5(1)(c)).

[37] The Minister of Agriculture, Fisheries and Food acts as the licensing authority for the scheme under Part II of the Food and Environmental Protection Act 1985

THE LEGAL ROLE OF THE NATIONAL RIVERS AUTHORITY IN RESPECT OF BATHING WATERS

The Water Act 1989 established the administrative division between the Water Service Public Limited Companies, with responsibility for water supply and sewage treatment, and the National Rivers Authority, a public body undertaking a range of functions concerned with the regulation of the aquatic environment. In particular, the National Rivers Authority is the competent authority for the enforcement of the Bathing Waters Directive in England and Wales, and as such the Authority is possessed of a range of legal powers and duties in relation to water quality. In the most general terms these legal powers and duties operate on three levels: first, as a broadly formulated strategic duty in relation to the aquatic environment including matters of water quality; second, specified monitoring and law enforcement duties in respect of the control of water pollution; and, third, the particular power to licence benign emissions into certain waters in accordance with the discharge consent system.

THE GENERAL ENVIRONMENTAL AND RECREATIONAL DUTIES

Amongst the various powers and duties of the National Rivers Authority which may be relevant to the quality of bathing waters a distinction is to be drawn between those functions which are general and cross functional in character, and those matters which are specifically related to the control of pollution. Functions of the latter kind will be considered later. The former consist of General Environmental and Recreational Duties imposed upon the Authority.

The General Environmental Duty stipulates that, amongst other matters, it is the duty of the Authority, *so far as is consistent with its functions*, to exercise any power conferred on it with respect to proposals as to further the conservation and enhancement of natural beauty and the conservation of flora and fauna; and to take into account any effect which the proposals would have on the beauty or amenity of any rural or urban area or on any flora, fauna, features, buildings, sites or objects.[38] Without prejudice to this, it is the duty of the Authority, *to such extent as it considers desirable* generally to promote the conservation and enhancement of the natural beauty and amenity of inland and coastal waters and of land associated with such waters, the conservation of flora and fauna which are dependent on an aquatic environment, and the use of such waters and land for recreational purposes[39].

The General Recreational Duty explicitly requires the Authority,[39] in formulating

---

[38] s.8(1) Water Act 1989. The same threefold environmental duty is imposed upon the Secretary of State, the Minister of Agriculture, Fisheries and Food, the Director of Water Services and every water undertaker, sewerage undertaker and internal drainage board *Ibid*. s.8(1) and (7).

[39] The same duty is imposed upon the Secretary of State, the Minister of Agriculture, the Director General of Water Services, water and sewerage undertakers and internal drainage boards *Ibid*. s.8(2)).

or considering any proposals, to have regard to the desirability of preserving for the public any freedom of access to certain open areas including the foreshore.[40] The duty requires the Authority, and other bodies including water and sewerage undertakers, to take such steps as are reasonably practicable and consistent with its functions, for securing that its rights to the use of water or land associated with water are exercised so as to ensure that the water or land is made available for recreational purposes and is so made available in the best manner.[41] It is the particular duty of the Authority, to such an extent as it considers desirable, generally to promote the use of inland and coastal waters and land associated with such waters for recreational purposes.[42]

CODES OF PRACTICE WITH RESPECT TO THE RECREATIONAL DUTY

The General Environmental and Recreational Duties are formulated in the most general language. More specific effect is given to these Duties by provision for Codes of Practice to be made to translate the general duties into more specific obligations. Hence the relevant Minister is empowered to approve any code of practice issued for the purpose of giving practical guidance to the Authority, or to water and sewerage undertakers, with respect to any of the matters in relation to which the General Environmental and Recreational duties are imposed, and promoting desirable practices by the Authority or such undertakers with respect to those matters.[43] A contravention of a code of practice is not of itself a contravention of the general environmental and recreational duties, but the Secretary of State and the Minister are under a duty to take into account whether these has been or is likely to be such contravention in determining when and how they should exercise powers under the 1989 Act in relation to the Authority or any water or sewerage undertaker.[44] No Ministerial order providing for a code of practice with respect to environmental and recreational duties is to be made unless there has been consultation with a range of bodies including the Authority, the Sports Council, the Sports Council for Wales, and such other water undertakers, sewerage undertakers and other persons as the relevant Minister considers it appropriate to consult.[45]

---

[40] *Ibid.* s.8(2).
[41] *Ibid.* s.8(3).
[42] *Ibid.* s.8(4).
[43] *Ibid.* s.10(1).
[44] *Ibid.* s.10(2). See s.20 in relation to enforcement orders against undertakers; and s.146 in respect of powers to issue directions to the Authority.
[45] *Ibid.* s.10(4). The "relevant Minister" means, in relation to the Authority, the Secretary of State or the Minister, and in relation to a water or sewerage undertaker the Secretary of State s.10(5).

THE CODE OR PRACTICE ON CONSERVATION, ACCESS AND RECREATION

The Ministerial power to make an order approving a code of practice with respect to environmental and recreational duties has been exercised under the Water and Sewerage (Conservation, Access and Recreation) (Code of Practice) Order (1989).[46] This Order constitutes formal approval of the Code of Practice on Conservation, Access and Recreation laid before Parliament in accordance with the 1989 Act.[47] The *Code* recognises that the bodies to whom it applies, the Authority and the water and sewerage undertakers, must operate with due regard to the proper performance of their statutory functions relating to abstraction, use and protection of water resources. Moreover, there may on occasions be a need to reconcile competing demands arising in relation to the performance of their duties in individual circumstances. Pertinently, however, the *Code* makes some observations with particular relevance to bathing waters in noting that the locations and lengths of proposed sea outfalls should be discussed with the appropriate conservation and planning bodies, so as to minimise construction damage and damage to marine ecology. Also, appropriate recreational bodies should be consulted in the planning of sea outfalls in view of their significance for bathing beaches, sub-aqua diving and angling.[48]

"CONTROLLED WATERS"

The National Rivers Authority has a range of responsibilities in relation to the aquatic environment, encompassing the control of pollution, water resources, flood defence, salmon and freshwater fisheries, and navigation, conservancy and harbour authority functions. The General Environmental and Recreational Duties apply to all the various functions of the Authority. In relation to the control of pollution alone there are some more particular powers and duties which may be of relevance to bathing waters.[49] In each case the powers and duties concerned are exerciseable in relation to "controlled waters" which is defined to encompass four subcategories: "relevant territorial waters", "coastal waters", "inland waters" and "ground waters".[50] In relation to sea bathing the first category is of greatest importance. *Relevant territorial waters* means waters which extend seaward for three nautical miles from the baselines from which the breadth of the territorial sea adjacent to England and Wales is measured.[51] This definition is subject to the power of the Secretary of State to provide by order that any particular area of territorial sea adjacent to England and

---

[46] SI 1989 No.1152.
[47] Required by s.10(3) of the Act..
[48] pp. 17 and 27.
[49] Under Chapter I (Control of Pollution) of Part III (Protection and Management of Rivers and Other Waters) of the Water Act 1989 (ss.103 to 124).
[50] s.103(1).
[51] s.103(1)(a).

Wales is to be treated as if it were an area of relevant territorial waters.[52] Hence, although the Authority is generally concerned with sea pollution to a distance of three miles from the low water mark,[53] there remains the possibility that its pollution control responsibilities could be extended beyond that distance if there were good reason for doing so.

WATER QUALITY CLASSIFICATIONS AND OBJECTIVES

The function of the Authority in relation to control of pollution allows for a strategic approach to the maintenance and improvement of water quality by the imposition of a system for the classification of the quality of waters.[54] Accordingly, it is provided that the Secretary of State may, in relation to any description of controlled waters, prescribe a system of classifying the quality of those waters according to certain criteria.[55] The criteria to be used may consist of general requirements as to the purpose for which the waters are to be suitable; specific requirements as to the substances that are to be present in or absent from the water and the concentrations of those substances; specific requirements as to other characteristics of those waters.[56] In accordance with this strategy the Surface Waters (Classification) Regulations 1989 [57] and the Surface Waters (Dangerous Substances) Regulations 1989 [58] have been made, but the formulation of a complete system for the classification of water quality awaits the outcome of the comprehensive survey taking place during 1990.[59]

The second element in strategic approach to the maintenance and improvement of water quality involves the establishment of a water quality classification in relation to each water. To achieve this the Secretary of State is empowered to serve a notice on the Authority specifying one or more of the prescribed water quality classifications and, in relation to each specified classification, a date.[60] When this has been done, the water quality objectives for particular waters will be the

---

[52] *Ibid.* s.103(5)(a).
[53] The Authority also has the power to regulate discharges of trade or sewage effluent from land in England and Wales, through a pipe, into the sea outside the seaward limits of controlled waters (*Ibid.* s.107(1)(c)(ii)) consequently long sea outfalls of three nautical miles in length or greater do not escape regulation for that reason.
[54] Previously systems for the classification of water quality had been employed as an administrative matter: see National Water Council: *River Water Quality, the Next Stage. Review of Consent Conditions*, 1978; and Department of the Environment: *Water Quality in England and Wales 1985*, 1986.
[55] *Ibid.* s.104(1).
[56] *Ibid.* s.104(2).
[57] SI 1989 No.1148.
[58] SI 1989 No.2286.
[59] National Rivers Authority *Annual Report 1989/90* (1990) p.20.
[60] *Ibid.* s.105(1).

satisfaction by those waters, on and at all times after the date specified, of the requirements for the classification set in relation to the waters.[61]

The third element in the strategic approach to maintenance and improvement of water quality concerns the relationship between the system of water quality objectives and other powers and obligations of the Secretary of State and the Authority in relation to water pollution. It is stated to be the duty of the Secretary of State and the Authority to exercise the powers conferred on them under the Water Act 1989, in relation to the control of pollution, in such manner as ensures, *so far as it is practicable by the exercise of those powers to do so*, that the water quality objectives specified for any waters are achieved at all times.[62] Accordingly, all the particular powers given to the Authority, such as the power to bring legal proceedings in relation to particular pollution incidents, are to be used with a particular purpose in mind: that of meeting and maintaining water quality objectives at all times.

A final observation on the system of Water Quality Classification and Water Quality Objectives concerns their relationship with the water quality requirements provided for by Community Directives. Where a water has been identified as falling within the Bathing Water Directive, necessarily, the water quality objective which is set for the water will be at least as strict as is required by the Directive. Similarly in relation to waters designated in relation to the Freshwater Fish or Shellfish Directives, the objectives which are set will reflect the use to which the waters are being put. The classification system and objectives for particular waters will also reflect other water Directives such as the Dangerous Substances Directive in establishing the parameters for the amounts of dangerous substances which are permitted to be present in the waters. In effect, therefore, the system of water quality classification and water quality objectives provides a refined formal mechanism for securing compliance with Directives.

OFFENCES OF POLLUTING CONTROLLED WATERS

On the level of controls upon particular polluting operations and incidents, the pollution of controlled waters is a criminal offence, in respect of which, proceedings may be brought by the Authority. In relation to bathing waters, perhaps the most relevant provisions are the offences which are committed where a person "causes or knowingly permits" [63] any poisonous, noxious or polluting matter or any solid waste matter to enter any controlled waters;[64] or any trade effluent or sewage effluent to be

---

[61] *Ibid.* s.105(2).
[62] *Ibid.* s.106(1).
[63] There is an extensive body of case law on the meaning of the phrase "causes or knowingly permits: generally see Howarth, *Water Pollution Law*, (1988) SS.4.05 to 4.07.
[64] *Ibid.* s.107(1)(a).

discharged into any controlled waters or from land in England and Wales, through a pipe, into the sea outside the seaward limits of controlled waters.[65]

In relation to the key problem of pollution of bathing waters caused by sewage effluent, some statutory definitions may be noted. *Effluent* is defined as any liquid, including particles of matter and other substances in suspension in the liquid.[66] *Sewage effluent* is stated to include any effluent from the sewage disposal or sewerage works of a sewerage undertaker but not to include surface water.[67] In relation to the offence of causing or knowingly permitting the discharge of sewage effluent into controlled waters a form of strict liability is imposed upon sewerage undertakers. This provides that where any sewage effluent is discharged into controlled waters or from land into waters beyond the seaward limits of controlled waters, and the undertaker did not cause or knowingly permit the discharge but was bound (either unconditionally or subject to conditions which were observed) to receive into the sewer or works matter included in the discharge, the undertaker shall be deemed to have caused the discharge.[68] In effect, therefore, a sewerage undertaker will not be able to argue that he has not caused pollution by sewage effluent in circumstances where the effluent is of a kind that he was bound to receive at his treatment works.

The initial year after privatisation has been allowed as an amnesty period during which the Authority has had time to gather the information required to support legal proceedings, and the sewage treatment works have had time to improve the quality of their effluent. Accordingly no prosecutions have been brought for this offence in relation to treatment works operating within their normal consent limits.[69] It is clear, however, that proceedings of this kind are imminent and that the Authority is determined to use the law to compel treatment works to improve their operations.[70] It is also clear that substantial increases have been seen in the amounts of fines that imposed by courts in relation to water pollution offences compared with the way in which this offence would previously have been regarded. Recently the maximum fine that could be imposed on summary conviction by a magistrates' court was increased from £2,000 to £20,000, as an alternative, or in addition, a convicted person may be sentenced to a period of imprisonment for a term of not exceeding three months.[71] On conviction on indictment, before the crown court, the

---

[65] *Ibid.* s.107(1)(c). "Outside the seaward limits of controlled waters" would generally mean beyond the three nautical mile limit, see above.
[66] *Ibid.* s.189(1).
[67] *Ibid.* s.124(1).
[68] *Ibid.* s.107(5).
[69] Prosecutions have, however, been brought for sewage treatment works exceeding absolute upper limits on discharge consents which are not subject to the same evidential requirements.
[70] See "Time is up for Water PLCs", *The Water Guardians* October 1990 p.1.
[71] s.107(6)(a) Water Act 1989, and see s.145 Environmental Protection Act 1990 (applicable from 1 January 1991).

corresponding penalties are imprisonment for a term not exceeding two years or to a fine of unlimited amount or both.[72] Moreover, there is evidence that the crown court is prepared to make far greater use of its punitive powers in relation to environmental offences. In the recent decision in *National Rivers Authority v Shell U.K.* [73] the polluting company were fined the sum of £1 million with an indication that, were it not for the previously good environmental record of the company, the fine would have been *substantially greater*.

DISCHARGE CONSENTS

The basic legal principle that the discharge of sewage effluent into controlled waters is a criminal offence is subject to a number of particular defences. Most pertinently in relation to the problem of pollution of bathing waters by sewage effluent, a defence is provided for so that a person will not be guilty of the offence in relation to any discharge which is made under an in accordance with, or as a result of any act or omission under and in accordance with, a discharge consent.[74] This is a reference to the power of the Authority to authorise certain emissions into the aquatic environment where such emissions are within acceptable limits in relation to the pollutants discharged and the nature and uses of the receiving waters. The detailed provisions relating to the granting of discharge consents are grouped together in Schedule 12 to the Water Act 1989.[75]

In general terms, Schedule 12 requires applications for discharge consents to be accompanied or supplemented by all such information as the Authority may reasonably require. Notice of an application is to be published by the Authority and copies of the application sent to every local authority or water undertaker within whose area the proposed discharge is to occur unless the discharge will have no appreciable effect on the waters into which the discharge is to be made. The Authority is then to consider any written representations or objections to the application made within a specified period. Publicity requirements may be circumvented where the Secretary of State is satisfied that disclosure of information about the discharge would be contrary to the public interest or would prejudice to an unreasonable degree some private interest by disclosing information about a trade secret. Consents may be granted subject to such conditions as the Authority may think fit and may include conditions such as to the nature, origin, composition, temperature, volume and rate of the discharges and as to the periods during which the discharges may be made. Provision is made for references of applications for

---

[72] s.107(6)(b) Water Act 1989.
[73] Liverpool Crown Court 23 February 1990, and reported in [1990] *Water Law* p.40.
[74] s.108(1)(a) Water Act 1989. A "discharge consent" refers to a consent provided under Chapter I of Part III of the Water Act 1989 (Concerned with the control of pollution) or under Part II of the Control of Pollution Act 1974 which amounted to the same thing under former legislation.
[75] Applicable under s.113(1) Water Act 1989.

discharge consents and appeals against determinations of the Authority to be made to the Secretary of State.[76]

The power to grant or decline applications for discharge consents, and to allow consents subject to conditions as to the levels of pollutants that may be discharged, is a profoundly useful mechanism for the selective authorisation of potentially polluting discharges into the aquatic environment. The basis upon which an application for a discharge consent is considered, or the review of an existing consent is undertaken, will be by reference to the capacity of the receiving waters to accept the effluent without an excessive deterioration in quality. In the past this may have been accomplished in a fairly rough and ready way, and there is some evidence of widely variable standard setting in respect of discharge consents for sewage treatment works.[77] For the future the link between the discharge consent system and the water classification system will become closer than previously and the water quality objective set for the receiving waters will, unavoidably, be a major determinant is assessing a permissible level of pollutant discharge. Accordingly if the receiving waters are bathing waters, or flow into bathing waters, the objective in setting discharge consent conditions will be to maintain the receiving waters at a standard which is at least as high as is required for waters used for bathing. Similarly, where the receiving waters are designated for another purpose the conditions contained in discharge consents will reflect the use to be made of the receiving waters.

FUTURE CONSIDERATIONS IN THE LAW RELATING TO BATHING WATERS

This survey has sought to draw attention to the key legal issues surrounding the Bathing Water Directive and its implementation. It is apparent, however, that significant difficulties remain with the implementation of the Directive and with the quality of bathing waters generally. From amongst the range of issues taken up for discussion by the House of Commons Environment Committee in its fourth report, *Pollution of Beaches*,[78] a number are likely to be of special importance. Most notably, however, comment is warranted on three issues of particular legal significance raised by the Committee: the designation of inland bathing waters; the legal position of water contact sports other than bathing; and the extent of the duty to warn bathers of dangers to which they are exposed.[79]

The Committee noted that, at present, no inland waters are designated under the terms of the Bathing Water Directive and recommended that the Authority should review inland waters used for water sports and bathing, with a view to appropriate designation. Whilst issue might be taken on the issue of whether the Directive allows

---

[76] Generally see Schedule 12 Water Act 1989; for changes to the discharge consent system which are presently under discussion see *Discharge Consent and Compliance Policy* NRA July 1990.
[77] Generally see *Ibid*.
[78] (1990) House of Commons Papers for Session 1989-90 12-I.
[79] See *Ibid*. paras.62 to 72.

for the "designation" of bathing waters, as opposed to the *identification* of those waters which, as a matter of fact, do come within the terms of the Directive, the absence of any inland waters from the list of waters within the Directive is remarkable. There would seem to be no explanation for this omission other than the fact that no inland waters satisfy the criterion that bathing in not prohibited and is traditionally practised by a large number of bathers. Given the duties of the Authority and other bodies to make inland waters available for recreation,[80] and the increasing popularity of inland waters as an amenity, the satisfaction of this criterion in the future is considerably more likely than in the past. If satisfied, it would be necessary for the waters concerned to meet the parameters provided for in the Bathing Water Directive, in addition to the water quality requirements of other Directives within which the waters fell. Accordingly, the water quality objectives for the waters would reflect the highest quality parameters determined by the conjunction of Directives applicable to the waters.

A second, related, issue is the position of water-contact sports such as sailing and windsurfing in respect of the Bathing Water Directive. Although the Directive makes reference to waters in which *bathing* is traditionally practised, it can be argued that if the objective of the Directive is to protect the environment and public health, as stated in the preamble, then no distinction should be drawn between customary bathing and other activities involving water contact which pose similar risks to the health of participants through contact with contaminated water.[81] Sidestepping the issue of whether waters in which water-contact activities take place are within the Directive, the Committee took the view that water quality objectives under the Water Act 1989 should ensure that a clean environment for water-contact sports is established and safeguarded. Whether or not such waters are "bathing waters" for the purposes of the Directive, it is to be hoped that the water classification system devised by the Secretary of State would reflect the public health risk involved to participants in water-contact sports by setting water quality objectives for waters in which those activities take place which are at least as strict as the requirements of the Directive.

Third, the Committee noted that persons are regularly bathing at beaches which do not meet the required standards, presumably at some risk to their health. This fact places local authorities in an invidious position in that they are responsible for matters of public health within their areas, but the posting of warning notices as to the danger of bathing in particular locations has serious implications for tourism. Despite this awkward dilemma, the view of the Committee was that there was an overriding public interest that persons making use of an amenity should be informed

---

[80] Under the General Recreational Duty discussed above.
[81] A contrary view has been expressed by Mr Ripa di Meana on behalf of the European Commission, who stated that water used solely for recreational sports other than bathing are not included within the scope of the Directive, and that their inclusion would require modification of the Directive by the unanimous decision of the Council (C303, 3 December 1990).

when regulatory standards imposed for their protection are breached. Consequently, the Committee expressed "great satisfaction" at the Ministerial announcement that a voluntary system of warnings by local authorities would be sought. In the event of that system not being acceptable to local authorities, however, it was thought that the National Rivers Authority should be given appropriate enforcement powers in relation to the placing of warning notices on unsatisfactory beaches. Although somewhat distasteful, the posting of warning notices alongside unsatisfactory beaches is, in law, a highly prudent course of action. Otherwise, it might be possible for a local authority to find itself subject to actions for negligence if bathers were to become seriously ill through bathing in unsatisfactory waters where the local authority had failed to issue warning of an identified danger to health. If only for fear of legal liability, it is to be hoped that local authorities will conform to this practice voluntarily without the need for enforcement powers to be vested in, and exercised by, the National Rivers Authority.

FOOTNOTES

[1] Directive concerning the quality of bathing water, 76/160/EEC of 10 December 1975.
[2] Art.189 Treaty of Rome 1957 Cmnd.7480, frequently referred to as the "EEC Treaty".
[3] *Ibid.* Art..69.
[4] *Ibid Art* 170.
[5] *Ibid Art* 171 .
[6] Case 96/81 [1982] ECR 1791.
[7] House of Commons Environment Committee, Fourth Report, *Pollution of Beaches,* (1990) House of Commons Paper 12-1.
[8] OJ 1990 C125/13
[9] [1990] *Water Law* 32.
[10] Contrast the powers to enact environmental legislation now provided for by the Single European Act 1987 which insets new Articles into the 1957 Treaty allowing for environmental legislation: see Articles 130R to 130T of the Treaty as amended.
[11] See House of Lords Select Committee on the European Communities, *Approximation of laws under Article 100 of the EEC Treaty,* 22nd Report Session 1977-78; and House of Lords' Select Committee on the European Communities, *Environmental Problems and the Treaty of Rome,* Session 1979-80.
[12] See Vandermeersch, "The Single European Act and the Environmental Policy of the European Economic Community", 12 *European Law Review* 407, at p.411.
[13] See, for example, the Directive on the quality of fresh waters needing protection or improvement in order to support fish life, 78/659/EEC.
[14] Art.1 Bathing Water Directive.
[15] Contrast the power of designation of waters provided for under the Directive on the quality of fresh water needing protection or improvement in order to support fish life, 78/659/EEC; and the Directive on the quality required for shellfish water, 79/923/EEC. [16] Haigh, *EEC Environmental Policy and Britain* (2nd Ed. 1987) p.63.

[17] Art.2(a) Bathing Water Directive.
[18] *Brinckman v. Mately* [1904] 2 Ch.313.
[19] s.231(1) (a) and (aa) Public Health Act 1936. The local authority concerned may exercise the powers provided in relation to public bathing in respect of any area of the sea which is outside its area but within 1,000 metres to seaward of any place where the low water mark is within or on the boundary of the authority's area (s. 17(1) Local Government (Miscellaneous Provisions) Act 1976).
[20] Haigh, *EEC Environmental Policy and Britain* (1987) p.66.
[21] Department of the Environment, Advice Notices of March 1977 and August 1979, discussed by Haigh, *Ibid.*
[22] Royal Commission on Environmental Pollution, Tenth Report, *Tackling Pollution Experience and Prospects,* (1984 Cm.9149) p.89.
[23] *Ibid.* para.4.69
[24] Department of the Environment, Pollution Paper No.22, *Controlling Pollution: Principles and Prospects* (1984) para.33.
[25] *Water Bulletin,* 23 November 1990 p.5.
[26] Art.3 Bathing Water Directive.
[27] Art.5(1).
[28] Art 4
[29] Art.8.
[30] Art.7.
[31] Art.6(1) and (2).
[32] Art.6(4)
[33] Directive on the quality required for shellfish waters, 79/923/EEC.
[34] Directive on pollution caused by certain dangerous substances discharged into the aquatic environment of the Community,76/464/EEC.
[35] Directive on the assessment of effects of certain public and private projects on the environment, 85/337/EEC.
[36] COM (89) 518 final.
[37] "Coastal waters", for these purposes, means waters within 3 nautical miles from any point on the coast measured from low water mark of ordinary spring tides (s.343(1) Public Health Act 1936).
[38] The Sea Fisheries Regulation Act 1966 gives powers to local sea fisheries committees to make byelaws. Specifically a local fisheries committee may, subject to regulations made by the Minister of Agriculture, Fisheries and Food, make byelaws for the purpose of prohibiting or regulating the deposit or discharge of any solid or liquid substance detrimental to sea fish or sea fishing (s.5(1) (c)).
[39] The Minister of Agriculture, Fisheries and Food acts as the licensing authority for the scheme under Part 11 of the Food and Environmental Protection Act 1985 concerning various operations relating to the deposit of waste at sea. In considering licence applications, the Minister is bound to have regard to the need to protect the marine environment, the living resources which it supports and human health. Also he is to have regard to the need to prevent interference with legitimate uses of the sea.
[40] s.8(1) Water Act 1989. The same threefold environmental duty is imposed upon the Secretary of State, the Minister of Agriculture, Fisheries and Food, the Director

of Water Services and every water undertaker, sewerage undertaker and internal drainage board *Ibid.* s.8(1) and (7).
[41] *Ibid.* s.8(4).
[42] The same duty is imposed upon the Secretary of State, the Minister of Agriculture, the Director General of Water Services, water and sewerage undertakers and internal drainage boards *Ibid.* s.8(2).
[43] *Ibid.* s.8(2).
[44] *Ibid.* s.8(3)
[45] *Ibid.* s.8(4)
[46] *Ibid.* s.10(1)
[47] *Ibid.* s.10(2). See s.20 in relation to enforcement orders against undertakers; and s.146 in respect of powers to issue directions to the Authority.
[48] *Ibid.* s.10(4). The "relevant Minister" means, in relation to the Authority, the Secretary of State or the Minister, and in relation to a water or sewerage undertaker the Secretary of State s. 10(5) .
[49] SI 1989 No.1152.
[50] Required by s.10(3) of the Act..
[51] pp 17 and 27.
[52] Under Chapter I (Control of Pollution) of Part 111 (Protection and Management of Rivers and Other Waters) of the Water Act 1989 (ss.103 to 124).
[53] s.103(1).
[54] s.103(1)(a)-
[55] *Ibid.* s.103(5)(a)
[56] The Authority also has the power to regulate discharges of trade or sewage effluent from land in England and Wales, through a pipe, into the sea outside the seaward limits of controlled waters *(Ibid.* s.107(1)(c)(ii)) consequently long sea outfalls of three nautical miles in length or greater do not escape regulation for that reason.
[57] Previously systems for the classification of water quality had been employed as an administrative matter: see National Water Council: *River Water Quality, the Next Stage. Review of Consent Conditions,* 1978; and Department of the Environment: *Water Quality in England and Wales 1985,* 1986.
[58] *Ibid.* s.104(1)
[59] *Ibid.* s.104(2).
[60] SI 1989 No.1148. 61 SI 1989 No.2286.
[62] National Rivers Authority *Annual Report 1989/90* (1990) p.20.
[63] *Ibid.* s.105(1)
[64] *Ibid.* s.105(2)
[65] *Ibid.* s.106(l)
[66] There is an extensive body of case law on the meaning of the phrase "causes or knowingly permits: generally see Howarth, *Water Pollution Law,* (1988) SS.4.05 to 4.07.
[67] *Ibid.* s.107(1)(a)
[68] *Ibid.* s.107(1) (c). "Outside the seaward limits of controlled waters" would generally mean beyond the three nautical mile limit, see above.
[69] *Ibid.* s.1 89(1).

[70] *Ibid.* s.124(1).
[71] Ibid.s.107(5)
[72] Prosecutions have, however, been brought for sewage treatment works exceeding absolute upper limits on discharge consents which are not subject to the same evidential requirements.
[73] See "Time is up for Water PLCs", *The Water Guardians* October 1990 p.1.
[74] s.107(6)(a) Water Act 1989, and see s.145 Environmental Protection Act 1990 (applicable from 1 January 1991).
[75] s.1 07(6)(b) Water Act 1989.
[76] Liverpool Crown Court 23 February 1990, and reported in [1990] *Water Law* p.40.
[77] s.108(1)(a)Water Act1989. A"discharge consent"refers to a consent provided under Chapter I of Part 111 of the Water Act 1989 (Concerned with the control of pollution) or under Part 11 of the Control of Pollution Act 1974 which amounted to the same thing under former legislation.
[78] Applicable under s.113(1) Water Act 1989.
[79] Generally see Schedule 12 Water Act 1989; for changes to the discharge consent system which are presently under discussion see *Discharge Consent and Compliance Policy* NRA July 1990.
[80] Generally see *Ibid.*
[81] (1990) House of Commons Papers for Session 1989-9012-1.
[82] See *Ibid.* paras.62 to 72.
[83] Under the General Recreational Duty discussed above.
[84] A contrary view has been expressed by Mr Ripa di Meana on behalf of the European Commission, who stated that water used solely for recreational sports other than bathing are not included within the scope of the Directive, and that their inclusion would require modification of the Directive by the unanimous decision of the Council (C303, 3 December 1990).

Chapter 6

MICROBIOLOGICAL ASPECTS AND POSSIBLE HEALTH RISKS OF RECREATIONAL WATER

Lorna Fewtrell and Frank Jones
Research Fellow and Director; Centre for Research into
Environment and Health, University of Wales, Lampeter SA48 7ED

INTRODUCTION

The association between water and disease has been recognised since ancient times. Until recently the majority of public concern has, not surprisingly, been related to the quality of our drinking water supplies. As water sports have become more varied and accessible, concern has increasingly been shown over the microbiological and chemical quality of recreational water, especially bathing beaches. In particular, emotion has been aroused in respect to health, aesthetics and the environment. Headlines such as "Bathers face infection from sea", "Britain fifth from bottom of Blue flag league" and "Delaying tactics on dirty beaches", have been seen in the national press.

Every day large amounts of sewage, agricultural waste and industrial effluent are discharged to UK surface waters. Although much of this waste will have been treated, the processes involved do not completely remove microbiological contaminants. Indeed it has been noted that the levels of bacteria and viruses in effluent from treated sewage can be appreciable (House of Commons Environment Committee, 1990). Depending upon the original source, a range of microorganisms present may be pathogenic. On ingestion or contact with a susceptible host these may cause infection and illness. Waters receiving these effluents will be microbiologically contaminated. Many of these receiving waters are extensively used for a variety of recreational activities, such as swimming, water skiing, canoeing and fishing.

Although in the UK the classical water-borne infections such as enteric fever and cholera are very rare, (and are generally only seen in travellers returning from countries with poor sanitation) in developed countries world-wide, epidemiological studies have identified a number of minor illnesses which have been associated with the recreational use of water (Stevenson, 1953; Dufour, 1982, 1984: Cabelli *et al.*, 1983; Philipp *et al.*, 1985; Seyfried *et al.*, 1985 a, b; DeWailley *et al.*, 1986; Ferley *et al.*, 1989; Balarajan, 1990; Jones *et al.*, 1991).

This paper details many of the pathogenic microorganisms which have been identified in UK recreational waters, and reports some of the incidents of infection which have been documented in the literature.

PATHOGENS

Many of the pathogens (excluding viruses) which have been isolated from UK surface waters are shown in Table 6.1. A large number of these microorganisms have other, more commonly recognised routes of transmission. If recreational water is suspected as a source of infection, retrospective sampling and analysis makes confirmation difficult. So, not surprisingly, some of the pathogens shown in Table 6.1 have not been directly implicated in recreational water infection, but their presence should not be dismissed. Environmental, as opposed to clinical, sampling and analysis often presents greater problems, especially in the isolation and identification of pathogenic organisms. In this regard, dilution and dispersion in receiving waters, the absence of routine monitoring for fresh waters and the indifference shown by laboratory services capable of carrying out such work, has resulted in a lack of data for many sites.

Some of the pathogens in Table 6.1 are dealt with in more detail. These have been highlighted because there are well documented cases of infection from recreational water or because they are particularly interesting.

Table 6.1 – Pathogens from UK recreational water.

Pathogen	Infection	Water type	Comments
Aeromonas spp.	Wound infection, Gastroenteritis	Fresh / marine	Autochthonous.
Campylobacter spp.	Enteritis	Fresh / marine	Contaminated food and milk the principal sources of infection.
Candida albicans	Dermatitis, Thrush	Fresh / marine	Mainly endogenous infections, although Brisou (1975) has noted an increase in Thrush infections amongst women holidaying at the seaside.
Clostridia spp.	Botulism, tetanus Gas gangrene, Gastroenteritis	Fresh / marine	Intestinal commensal in some mammals including man. Most human infection through soil contaminated wound infection. Food poisoning toxins. Some water-related bird deaths due to botulism.

Cryptosporidia spp.	Gastroenteritis	Fresh (marine)	Gastrointestinal infection known as cryptosporidiosis. Common livestock parasite. There have been a number of drinking water outbreaks in the UK. Also a recreational water outbreak due to a contaminated swimming pool in Yorkshire.
Escherichia coli	Gastroenteritis	Fresh / marine	Used as an indicator of faecal contamination. A few specialised enteropathogenic species, associated with food poisoning and infant infections.
Giardia spp.	Gastroenteritis	Fresh (marine)	Infection known as giardiasis. Host vectors include beaver and deer.
Legionella spp.	Legionellosis	Fresh	Legionellosis - has two distinct clinical presentations: Legionnaires' disease, an acute pneumonia with a low attack rate, but relatively high fatality rate, and Pontiac fever, a mild febrile illness. Legionellae have a wide distribution in the environment, including cooling towers and plumbing systems. Intracellular association with amoebae.
Leptospira spp.	Leptospirosis	Fresh	Potentially fatal. Leptospirosis is a notifiable disease in England and Wales. Rodent and cattle vectors.
Naegleria spp.	Primary amoebic meningoencephalitis (PAM)	Fresh	PAM very rare in UK. Thermophilic organisms.
Plesiomonas spp.	Gastroenteritis Meningitis Cellulitis	Fresh / estuarine occasionally marine	Gastroenteritis - controversial
Pseudomonas spp.	Follicular dermatitis Otitis externa	Fresh / marine	Often associated with swimming pool infections.
Salmonella spp.	Enteric fever Gastroenteritis	Fresh / marine	Enteric fever (typhoid and paratyphoid) rare. Gastroenteritis (salmonellosis) more commonly associated with food poisoning.
Shigella spp.	Bacillery dysentery	Fresh / marine	Generally person to person transmission.
Staphylococcus spp.	Soft tissue infections Bacteraemia	Fresh / marine	Human source, indicates bather concentration pollution.
Vibrio spp.	Cholera Wound infections	Predominantly marine and estuarine	Cholera very rare in UK.
Yersinia spp.	Gastroenteritis	Fresh / marine	Zoonotic agents.

AEROMONAS SPP.

In the literature most water associated aeromonad infections have been due to *Aeromonas hydrophila*, although Delbeke *et al.* (1985) has reported two infections due to *A. sobria*. *A. hydrophila* is widely distributed, and is commonly found in freshwater streams and lakes as well as brackish and saline water. Indeed, it is considered to be an autochthonous (indigenous) inhabitant of the aquatic environment (Chowdhury *et al.*, 1990).

In the UK, there have been no documented cases of recreational water aeromonad infection, however, world-wide there have been a number of reports. The cases have mainly followed traumatic injury in contaminated water. Hanson *et al.* (1977) reported a case of aeromonad cellulitis following wounds sustained whilst swimming. *A. hydrophila* has also been isolated from endotracheal aspirates of patients suffering from aspirational pneumonia after near drowning (Simon and Von Graevenitz, 1969; Reines and Cook, 1981).

CAMPYLOBACTER SPP.

In Britain, campylobacters are the most frequently identified cause of acute bacterial enteritis (Skirrow, 1986). Campylobacter enteritis is manifest by a fever of 40°C, followed by severe abdominal cramp, nausea and diarrhoea. The diarrhoea is either watery and profuse or slimy and often contains blood.

The principal sources of human infection are associated with contaminated food (especially poultry) and milk. There have, however, been a number of reports of infection through the consumption of contaminated water (Mentzing, 1981; Taylor *et al.*, 1983; Aho *et al.*, 1989) and one following a fall into sewage (Sumathipala and Morrison, 1983). Campylobacters have been isolated from a range of marine and freshwater sites, including mountain streams, lakes and ponds (Carter *et al.*, 1987; Jones *et al.*, 1990).

Skirrow (1987) demonstrated a peak of campylobacter enteritis infection in young males during the summer months. It was suggested that this may partly be explained by the large number of young men taking part in watersport activities and using camp sites without properly treated water. The organism is very sensitive to disinfectants such as chlorine and, compared to salmonellae, has a lower minimal infecting dose.

CRYPTOSPORIDIUM SPP.

Cryptosporidia are protozoan parasites with a complex lifecycle. They have only recently been recognized as human pathogens (Casemore *et al.*, 1985) but are now known to be an important cause of gastroenteritis (cryptosporidiosis). The symptoms include stomach pains, fever, vomiting and diarrhoea. The illness usually abates within three weeks, but certain vulnerable groups (such as children, the elderly and immunocompromised) may be more severely affected.

Cryptosporidia are widespread in the environment. In the UK, most outbreaks of cryptosporidiosis have been limited to drinking water supplies (Oxfordshire and Swindon 1989, and North Humberside and Hull 1990). However, at least 70 people were infected from a swimming pool in Yorkshire, which was found to be heavily contaminated with *Cryptosporidium* oocysts (Galbraith, 1989; Barer and Wright, 1990). Research in the United States has also suggested an association between cryptosporidiosis and the recreational use of contaminated natural surface water (Gallagher *et al.*, 1989).

GIARDIA SPP.

Infection with pathogenic *Giardia* spp. leads to the gastrointestinal disease known as giardiasis. Infection results in a dense coating of parasites on the epithelial cells of the duodenum, this interferes with nutrient adsorption and is clinically manifest as explosive, watery foul diarrhoea often accompanied by nausea and vomiting. The acute stage of the illness usually lasts for 3 to 4 days, but may persist for months. The infective dose is thought to be low (Rendtoff, 1954).

There have been reports of giardia infection, mainly from the United States, following the consumption of untreated stream water (Wright, 75; Barbour *et al.*, 1976). In the UK, the only documented incident to date was in Bristol, apparently due to the contamination of a municipal water supply (Jephcott *et al.*, 1986).

In the USA, *Giardia lamblia* is one of the most commonly reported causes of water-borne illness (Wolfe, 1979; Craun, 1988). A number of surveys of giardia cyst levels in various waters have shown that 26-43% of the waters were contaminated with cysts, ranging in concentrations from 0.3 to 100 cysts/100l (Rose *et al.*, 1991) In Britain, recorded water-borne giardiasis is less common. According to Duerden *et al.* (1990) Giardia is the most commonly reported parasite in this country. The main route of transmission to humans is probably person to person spread, but other animals such as deer and beaver have been implicated (Davis and Hibler, 1979).

LEPTOSPIRA SPP.

Infections caused by pathogenic leptospires have been given a variety of common names, such as Swineherd's disease (*L. pomona*), Stuttgart disease (*L. canicola*) and Weil's disease (*L. icterohaemorrhagiae*). Collectively these infections are known as leptospirosis. In the area of freshwater recreation, leptospirosis (especially Weil's disease) is the most commonly known infection and almost certainly the most feared.

Infection is either through contact with infected urine (usually from rodents or cattle) or indirectly via urine contaminated freshwater or wet soil. The virulent leptospires gain entry through cuts and abrasions and also through the mucosal surfaces of the mouth, nose and conjunctiva. Leptospirosis is a potentially fatal disease, the severity of the infection may vary from hepatorenal failure and meningitis, associated with classic Weil's disease, to a milder flu-like illness with severe headache. Leptospirosis is a notifiable disease. In the period between

1980 – 1989, there were 78 reported cases of water related leptospiral infection from a total of 533 cases (with 25 fatalities over that period). Due to the non-specific presentation of leptospirosis, it is almost certain that there is underdiagnosis of the disease, with mild cases probably being dismissed as flu. However it is unlikely that a significant number of severe cases are missed (Ferguson, 1990).

Rodents are considered to be the main environmental source of *Leptospira icterohaemorrhagiae*, dogs of *L. canicola* and cattle and sheep of *L. hardjo*, although this is only a general rule. It is obviously impractical to eradicate all the potential sources of leptospiral infection. As the incidence of the disease is relatively low, the introduction of sensible precautions for high risk user groups is thought to be the best means of achieving protection. Sporting bodies such as the Royal Yachting Association, and many local authorities now produce information sheets designed to raise public awareness of the disease and encourage precautionary measures (such as covering cuts, and showering after immersion) to be taken. Such simple preventative measures used by sewer workers in the UK appear to have produced favourable results.

NAEGLERIA SPP.

The amoeba *Naegleria fowleri*, following invasion of the brain and meninges via the nasal mucosa and olfactory tissue, can cause the disease Primary amoebic meningoencephalitis (PAM) in humans. This disease, which is almost invariably fatal, is exceedingly rare in Britain. Symmers (1969) recorded two possible retrospective cases in the UK, one of which was thought to be attributable to a near drowning incident. The first definite incident of PAM in this country was reported by Apley *et al.* (1970). A young boy died several days after splashing about in a warm puddle, *N. fowleri* were isolated from his cerebrospinal fluid. In 1978 a girl died from PAM, 11 days after swimming in a pool fed by thermal spring water (Cain *et al.*, 1981).

N. fowleri is a thermophilic microorganism, its optimum growth is between the temperatures 37 - 45°C (Cain *et al.*, 1981). For this reason most reported cases of PAM occur in areas with a warm climate. *N. fowleri* is far more common in areas of the USA where water temperatures are greater than those in the UK, however, despite its widespread distribution in the States there have been very few cases.

Since the organism gains entry to the body via the olfactory tissue extended bottom swimming and diving are thought to facilitate the entry of the amoebae into the nasal passages possibly predisposing the individual to infection (Jamieson and Anderson, 1973; Wellings *et al.*, 1977).

SHIGELLA SPP.

There are four *Shigella* spp. which cause bacillery dysentery (or shigellosis). The severity of the infection varies depending on the species involved, in the following declining order of severity: *S. dysenteriae*, *S. flexneri*, *S. boydii* and *S. sonnei*. Infection with *S. dysenteriae* results in a severe illness in which there is a sudden onset of abdominal pain and sometimes convulsion. Stools are small, frequent and are composed mainly of fresh blood, pus and mucus. In contrast, *S. sonnei* shigellosis is usually confined to moderately severe diarrhoea lasting for several days.

Shigellosis is generally spread by the faecal-oral route, usually by hand to mouth, although there are several well documented cases of infection in the United States following contact with contaminated recreational water. Rosenberg *et al.* (1976) reported an incident where 31 of 45 cases of *S. sonnei* infection could be traced to swimming in a stretch of the Mississippi river. Swimming and illness showed a significant correlation ($p < 0.001$), and *S. sonnei* was isolated from the river water. In a second incident, Makintubee *et al.* (1987) linked an outbreak of shigellosis with bathing in a lake. A survey of lake visitors demonstrated that swimmers were 50% more likely to suffer from shigellosis than those who did not swim. The third reported outbreak was also associated with lake bathing, 68 people had diarrhoea within one week of swimming at the lake, 33 of these cases were culture confirmed as shigellosis (Sorvillo *et al.*,1988).

VIBRIO SPP.

Eleven species of *Vibrio* have been identified as human pathogens. Historically *Vibrio cholerae* 01, the causative agent of cholera, has been of the greatest interest to clinicians and microbiologists, however, in developed countries cholera is now rare and is generally only seen in travellers returning from countries with poor sanitation. Other *Vibrio* spp. have been associated with a range a clinical symptoms including severe gastrointestinal infections (in some cases mimicking cholera), bacteraemia, wound and ear infections.

In the UK pathogenic vibrio infections are relatively infrequent, especially when compared to other developed countries such as the United States. However, if public health laboratories use only routine enteric media for the isolation of gastroenteritis agents, it is possible that many cases of vibrio enteritis are overlooked (Blake *et al.*, 1980).

World-wide there have been a number of reports of vibrio wound infection following contact with contaminated recreational water. In America, a man developed leg gangrene and endotoxic shock due to vibrio infection, following an injury whilst bathing in coastal waters (Roland, 1970). Olsen (1978) reported two cases of ear infection due to *Vibrio parahaemolyticus* following seawater exposure. In the UK, Ryan (1976) reported three cases of vibrio infection, including a case of otitis externa, obtained along the Devon-Dorset coast.

Vibrio gastroenteritis is often associated with the consumption of raw seafood, in this country, Hooper *et al.* (1974) reported a case of gastroenteritis caused by *V. parahaemolyticus*, due to the consumption of locally produced crab meat.

VIRAL INFECTIONS

There are potentially a large number of pathogenic viruses which may be present in recreational water, these are shown in Table 6.2.

Due to the wide range of disease syndromes produced by viruses, and the fact that many are difficult to detect using conventional methods, ascribing viruses to cases of water-borne illness is far from easy. Nevertheless, there have been a number of water-borne recreational cases reported in the literature.

An outbreak of gastrointestinal illness occurred in visitors to a recreational park in the United States. A significant relationship was established between illness and swimming in the park lake. Norwalk virus was identified as the aetiological agent (Baron *et al.*, 1982). Hawley *et al.* (1973) and Denis *et al.* (1974) reported cases of minor disease outbreaks following freshwater contact, these were ascribed to Coxsackievirus A and B. In the US in 1969 there was an outbreak of Hepatitis A which was attributed to the consumption of polluted lake water during recreation (Bryan *et al.*, 1974), and in this country a case of Hepatitis A associated with swimming in Bristol City Docks was identified (Philipp *et al.*, 1989).

The minimum infective dose in a susceptible individual may be as low as one viral particle. Any water body that receives sewage will contain enteric viruses, in a study of one inland river used for recreation over 80% of samples taken were found to contain viruses, in numbers ranging from 3 - 640 plaque forming units per 10 litres (Morris, 1989).

Table 6.2 – Human viruses potentially present in recreational water. (adapted from Rao and Melnick, 1986).

Virus group	Infection and disease symptoms in humans
Poliovirus	Paralysis, meningitis, fever.
Echovirus	Meningitis, respiratory disease, rash, fever, gastroenteritis.
Coxsackievirus A	Herpangina, respiratory disease, meningitis, fever.
Coxsackievirus B	Myocarditis, rash, fever, meningitis, respiratory disease, pleurodynia.
Enterovirus types 68 - 71	Meningitis, encephalitis, respiratory disease, rash, acute haemorrhagic conjunctivitis, fever.
Enterovirus 72	Infectious hepatitis.
Norwalk virus	Epidemic vomiting and diarrhoea, fever.
Rotavirus	Gastroenteritis, diarrhoea.
Adenovirus	Respiratory disease, conjunctivitis, gastroenteritis.

BLUE GREEN ALGAE

Blue green algae or cyanobacteria are potentially an important microbiological cause of water-associated illness in UK recreational waters. Problems arise following contact with or ingestion of water containing cyanobacterial toxins. Certain cyanobacteria are capable of producing a number of toxins, including hepatotoxins (such as the microcystins) and neurotoxins. In animals, the ingestion of microcystins has been shown to cause circulatory shock and rapid, fatal liver damage (Codd, 1984).

Although cases of animal deaths following cyanobacteria ingestion have been known for some time, the issue hit the headlines in Britain following the death of several dogs and 20 sheep (Edney, 1990; NRA, 1990) at Rutland Water in Leicestershire. Warning notices were promptly erected around the reservoir, banning the recreational use of the water. To date, there have been no human deaths attributed to blue green algae, although recently the first UK cases of suspected cyanobacterial poisoning were reported (Turner *et al.*, 1990). Two army recruits were admitted to hospital with an unusual form of pneumonia, 4-5 days after participating in a canoeing exercise on a lake containing *Microcystis aeruginosa*. Following treatment they both made a complete recovery.

INDICATORS

Looking at the range of pathogens identified in Tables 6.1 and 6.2, it might seem logical to monitor levels of these pathogens to assess possible health risks. In fact, the majority of monitoring is confined to non-pathogenic indicator bacteria, with only occasional samples examined for pathogens such as salmonellae and enterovirus.

Escherichia coli and faecal streptococci, which are abundant in human and animal excreta, are customarily used as indicators of faecal contamination. Recent studies have shown that *E. coli* is more sensitive than enterococci to natural waters (Pettibone *et al.*, 1987; Barcina *et al.*, 1990). However the issue is complicated taxonomically as far as the faecal streptococci and enterococci are concerned (Pourcher *et al.*, 1991). It has also been reported that *Streptococcus bovis* dies off much more rapidly in water than the enterococci (Geldreich and Kenner, 1969). This may explain contradictory evidence since it has been custom, by some, to regard *Strep. bovis* and *Strep. equinus* as one species and combine them in the enterococci as one group.

Geldreich and Kenner (1969) proposed a faecal coliform/faecal streptococci ratio to identify the origin of animal or human faecal pollution of surface waters. This approach has been highly controversial. However, because the proportions of the various streptococcal species are variable and not the same in human and animal faeces (Ratkowski and Sjogren, 1967) streptococcal identification rather than enumeration in water samples has been proposed (Pourcher *et al.*, 1991). With intensive animal husbandry there may be a need to pinpoint such sources of faecal pollution to water courses. This may be even more essential due to the zoonotic nature of some of the pathogens that may be present in such waters.

In the UK, the principle legislation covering recreational waters is the EC Bathing Water Directive (76/160/EEC). This legal document sets down the microbiological standards which should be met by all designated or identified bathing waters. Designated sites are sampled at least fortnightly during the summer bathing season and, during that time, should conform to maximum total and faecal coliform levels (for 'I' (mandatory) values on a 95 percentile basis and for 'G'(guide) values on an 80 percentile basis). Monitoring for faecal streptococci, salmonellae and enteroviruses is discretionary.

Quality standards for recreational waters are seldom based upon firm epidemiological evidence and hence standard systems are not without their own problems. Consequently a range of different standard systems are in use with considerable discussion centred upon their epidemiological validity. The relative stringency of these standards has been examined by Kay (1988) and Kay et al. (1989). However, it is possible to rank the relative stringency of these standards and show that the EC Directive mandatory level is the least stringent of the major international standards, whilst the guide, or recommended, level is the most demanding.

DISCUSSION

Most of the natural waters in use for bathing and other recreational purposes contain concentrations of faecal indicator organisms, and a varied range of pathogens. In the UK, this is due to a number of factors including: the practice of disposing of raw, partially treated, sewage effluent to surface waters; agricultural run off and the presence of naturally occurring pathogens in fresh and marine waters. It is unavoidable, therefore, that in many situations water recreators when bathing or undertaking watersports are exposing themselves to infection by a variety of harmful organisms. It may well be that more effective tertiary sewage treatment is used to reduce microoganisms to minimal or nil concentration. The use of modern microfiltration techniques and/or the application of ultraviolet radiation, which are both environmentally friendly, might offer the best way forward in treating discharges to recognised recreational waters. Chemical disinfection and the more novel application of ultrasonics are other alternative treatment options.

The legislation which currently exists in European Council Directive form in order to protect the environment and public health, surprisingly does not apply to freshwater sites used for bathing or recreation in the United Kingdom. Even though some health incidents have been recorded (Philipp et al., 1985) and other relevant scientific literature published (Jones and Godfree, 1989), the data necessary to draw firm decisions to advise managers and users of recreational waters is conspicuous by its near total absence. This situation is compounded further by the lack of freshwater epidemiological studies in the UK. Environmental managers thus face a problem when they seek to enforce standard parameters which are difficult to justify on scientific evidence.

The advent of new and more rapid microbiology techniques such as gene probes, impedimetry, mono-clonal antibodies, enzymatic methods and others, together with the increased awareness of recreators and government organisations will result in increased attention on the use and monitoring of recreational waters. It is essential to rectify the total lack of robust epidemiological data to assess potential risks to health in using waters for these activities. Indeed, future health and safety legislation and/or legal directives may well trigger scientific health investigations to safeguard those offering and those using water sport facilities.

More attention also need to be focussed on identifying and setting microbial standards for freshwaters in particular. How long can bacterial indicators continue to be used to assess viral survival in receiving waters? While viral analysis of environmental water samples is becoming more available, the use of F_2 bacteriophage may offer an acceptable alternative that is also cheap and easy to use. Decisions should be made in the light of present and future knowledge as to whether the same standards should be applied to all types of receiving waters and for all types of water recreation including bathing. Until such matters are dealt with, the environmental water recreational manager will continue to be placed in an invidious position making judgements on restricted guidance and advice. Whilst the recreators will have to rely on common sense in judging the the risks taken in using a specific facility for enjoyment.

REFERENCES

Aho, M., Kurki, M., Rautelin, H. and Kosunen, T.U. (1989) Waterborne outbreak of campylobacter enteritis after outdoors infantry drill in Utti, Finland. *Epidemiology and Infection 103*, 133-141.

Apley, J., Clarke, S.K.R., Roome, A.P.C.H., Sandry, S.A., Saygi, G., Silk, B. and Warhurst, D.C. (1970) Primary amoebic meningoencephalitis in Britain. *British Medical Journal i*, 596-599.

Balarajan, R. (1990) Health risks associated with bathing in the sea: results of a pilot study in Langland Bay. Guildford, University of Surrey.

Barbour, A.G., Nichols, C.R. and Fukushima, T. (1976) An outbreak of giardiasis in a group of campers. *American Journal of Tropical Medicine and Hygiene 25(3)*, 384-389.

Barcina, I., Gonzales, J.M., Iriberri, J. and Egea, L. (1990) Survival strategy of *Escherichia coli* and *Enterococcus faecalis* in illuminated fresh and marine systems. *Journal of Applied Bacteriology 68*, 189-198.

Barer, M.R. and Wright, A.E. (1990) *Cryptosporidium* and water. *Letters in Applied Microbiology 11*, 271-277.

Baron, R.C., Murphy, F.D., Greenberg, H.B., Davis, C.E., Bergman, D.J., Gary, G.W., Hughes, J.M. and Schonberger, L.B. (1982) Norwalk gastrointestinal illness.

An outbreak associated with swimming in a recreational lake and secondary person-to-person transmission. *American Journal of Epidemiology 115(2)*, 163-172.

Blake, P.A., Weaver, R.E. and Hollis, D.G. (1980) Diseases of humans (other than cholera) caused by vibrios. *Annual Review of Microbiology 34*, 341-367.

Brisou, J. (1975) Yeasts and fungi in marine environments. *Bulletin of the Society of French Mycology and Medicine 4*, 159-162.

Bryan, J.A., Lehmann, J.D., Setiady, I.F. and Hatch, M.H. (1974) An outbreak of Hepatitis A associated with recreational lake water. *American Journal of Epidemiology 99(2)*, 145-154.

Cabelli, V.J., Dufour, A.P., McCabe, L.J. and Levin, M.A. (1983) Marine recreational water quality criterion consistent with indicator concepts and risk analysis. *Journal of the Water Pollution Control Federation 55(10)*, 1306-1314.

Cain, A.R.R., Wiley, P.F., Brownell, B. and Warhurst, D.C. (1981) Primary amoebic meningoencephalitis. *Archives of Disease in Childhood 56*, 140-143.

Carter, A.M., Pacha, R.E., Clark, G.W. and Williams, E.A. (1987) Seasonal occurrence of *Campylobacter* species in surface waters, and their correlation with standard indicator bacteria. *Applied and Environmental Microbiology 53(3)*, 523-526.

Casemore, D.P., Sands, R.L. and Curry, A. (1985) Cryptosporidium species a 'new' human pathogen. *Journal of Clinical Pathology 38*, 1321-1336.

Chowdhury, M.A.R., Yamanaka, H., Miyoshi, S.I. and Shinoda, S. (1990) Ecology of mesophilic *Aeromonas* spp. in aquatic environments of a temperate region and relationship with some biotic and abiotic environmental parameters. *Zentralblatt fur Bakteriologie und Hygiene 190*, 344-356.

Codd, G.A. (1984) Toxins of freshwater cyanobacteria. *Microbiological Sciences 1(2)* 48-52.

Craun, G.F. (1988) Surface water supplies and health. *Journal of the American Water Works Association 80(2)*, 40-52.

Davis, R.B. and Hibler, C.P. (1979) Animal reservoirs and cross-species transmission of giardia. In: *Waterborne transmission of giardiasis* EPA-600/9-79-001 pp. 104-126.

Delbeke, E., Demarcq, M.J., Roubin, C. and Baleux, B. (1985) Contamination aquatique de plaies *Aeromonas sobria* apres bain de riviere. *La Presse Medicale 14(23)*, 1292.

Denis, F.A., Blanchovin, E., DeLignieres, A. and Flamen, P. (1974) Coxsackie A_{16} infection from lake water. *Journal of the American Medical Association 228*, 1370-1371.

DeWailley, E., Poirier, C. and Meyer, F.M. (1986) Health hazards associated with windsurfing on polluted waters. *American Journal of Public Health 76*, 690-691.

Duerden, B.I., Reid, T.M.S., Jewsbury, J.M. and Turk, D.C. (1990) Giardiasis. In: *Microbial and Parasitic Infection*. Edward Arnold, London. pp. 144.

Dufour, A.P. (1982) Fresh recreational water quality and swimming-associated illness. In: *Second National Symposium on Municipal Wastewater Disinfection*. Orlando, Florida.

Dufour, A.P. (1984) Bacterial indicators of recreational water quality. *Canadian Journal of Public Health 75*, 49-56.

Edney, A.T.B. (1990) Toxic algae. *The Veterinary Record 27.10.90*, 434-435.

Ferguson, I.R. (1990) Human Leptospirosis. *State Veterinary Journal 44 (125)*, 131-144.

Ferley, J.P., Zimrou, D., Balducci, F., Baleux, B., Fera, P., Larbuigt, G., Jacq, E., Mossonnier, B., Blineau, A. and Boudot, J. (1989) Epidemiological significance of microbiological pollution criteria for river recreational waters. *International Journal of Epidemiology 18(1)*, 198-205.

Galbraith, N.S. (1989) Cryptosporidiosis: another source. *British Medical Journal 298*, 276-277.

Gallagher, M.M., Herdon, J.L., Nims, L.J., Sterling, C.R., Grabowski, D.J. and Hull, H.H. (1989) Cryptosporidiosis and surface waters. *American Journal of Public Health 79*, 39-42.

Geldreich, E.E. and Kenner, B.A. (1969) Concepts of faecal pollution in stream pollution. *Journal of Water Pollution Control Federation 41*, R335-R352.

Hanson, P.G., Standridge, J., Jarret, F. and Maki, D.G. (1977) Freshwater wound infection due to *Aeromonas hydrophila.*. *Journal of the American Medical Association 238(10)*, 1053-1054.

Hawley, H.B., Morin, D.T., Geraghty, M.E., Tomkow, J. and Phillips, C.A. (1973) Coxsackievirus B epidemic at a boys' summer camp. Isolation of virus from swimming water. *Journal of the American Medical Association 226*, 33-36.

Hooper, W.L., Barrow, G.I. and McNab, D.J.N. (1974) *Vibrio parahaemolyticus* food poisoning in Britain. *Lancet i*, 1100.

House of Commons Environment Committee (1990) *Pollution of beaches*. London: HMSO, fourth report of the Environment Committee.

Jamieson, A. and Anderson, K. (1973) A method for the isolation of *Naegleria* species from water samples. *Pathology 5*, 55-58.

Jephcott, A.E., Begg, N.T. and Baker, I.A. (1986) Outbreak of Giardiasis associated with mains water in the United Kingdom. *Lancet i*, 730-732.

Jones, F. and Godfree, A.F. (1989) Recreational and amenity aspects of surface waters: the public health implications. *Water Science and Technology 21(3)*, 137-142.

Jones, F., Kay, D., Stanwell-Smith, R. and Wyer, M. (1991) Results of the first pilot-scale controlled cohort epidemiological investigation into the possible health effects of bathing in seawater at Langland Bay, Swansea. *Journal of the Institution of Water and Environmental Management 5(1)*, 91-98.

Jones, K., Betaieb, M. and Telford, D.R. (1990) Thermophilic campylobacters in surface waters around Lancaster, UK: negative correlation with campylobacter infections in the community. *Journal of Applied Bacteriology 69*, 758-764.

Kay, D. (1988) Coastal bathing water quality: the application of water quality standards to Welsh beaches. *Applied Geography 8*, 117-134.

Kay, D., Wyer, M.D., McDonald, A.T. and Woods, N. (1989) *The application of water quality standards to UK bathing waters*. Research report for a WRC funded project.

Makintubee, S., Mallonee, J. and Istre, G.R. (1987) Shigellosis outbreak associated with swimming. *American Journal of Public Health 77(2)*, 166-168.

Mentzing, L.O. (1981) Waterborne outbreaks of campylobacter in Sweden. *Lancet ii*, 352-354.

Morris, R. (1989) *Microbiological quality of an inland surface water used for recreational purposes*. In Wheeler, D., Richardson, M.L. and Bridges, J. (Eds.):

Watershed 1989 - the future for water quality in Europe. Pergamon Press, Oxford. pp.353-356.

NRA (1990) Toxic blue-green algae. The report of the National Rivers Authority. *Water Quality Series No. 2*. National Rivers Authority, London.

Olsen, H. (1978) *Vibrio parahaemolyticus* isolated from discharge from the ear in two patients exposed to sea water. *Acta Pathologica, Microbiologica et Immunologica Scandanavica Sect. B 86*, 247-248.

Pettibone, G.W., Sullivan, S.A. and Shiaris, M.P. (1987) Comparative survival of fecal indicator bacteria in estuarine water. *Applied and Environmental Microbiology 53*, 1241-1245.

Philipp, R., Evans, E.J., Hughes, A.O., Grisdale, S.K., Enticott, R.G. and Jephcott, A.E. (1985) Health risks of snorkel swimming in untreated water. *International Journal of Epidemiology 14(4)*, 624-627.

Philipp, R., Waitkins, S., Caul, O., Roome, A., McMahon, S. and Enticott, R. (1989) Leptospiral and Hepatitis A antibodies amongst windsurfers and waterskiers in Bristol city docks. *Public Health 103*, 123-129.

Pourcher, A.M., Devriese, L.A., Hernandez, J.F. and Delattre, J.M. (1991) Enumeration by a minaturized method of *Escherichia coli*, *Streptococcus bovis* and enterococci as indicators of the origin of faecal pollution of waters. *Journal of Applied Bacteriology 70*, 525-530.

Rao, V.C. and Melnick, J.L. (1986) Environmental Virology. *Aspects of Microbiology 13*. Van Nostrand Reinhold (UK).

Ratkowski, A.A. and Sjogren, R.E. (1987) Streptococcal population profiles as indicators of water quality. *Water, Air and Soil Pollution 34*, 273-284.

Reines, H.D. and Cook, F. (1981) Pneumonia and bacteremia due to *Aeromonas hydrophila Chest 30*, 264-267.

Rendtorff, R.C. (1954) The experimental transmission of human intestinal protozoan parasites. II. *Giardia lamblia* cysts given in capsules. *American Journal of Hygiene 59(2)*, 209-220.

Roland, F.P. (1970) Leg gangrene and endotoxin shock due to *Vibrio parahaemolyticus* - an infection acquired in New England coastal waters. *The New England Journal of Medicine 283(23)*, 1306.

Rose, J.B., Haas, C.N. and Regli, S. (1991) Risk assessment and control of waterborne giardiasis. *American Journal of Public Health 81*, 709-713.

Rosenberg, M.L., Hazlet, K.K., Schaefer, J., Wells, J.G. and Pruneda, R.C. (1976) Shigellosis from swimming. *Journal of the American Medical Association 236*, 1849-1852.

Ryan, W.J. (1976) Marine vibrios associated with superficial septic lesions. *Journal of Clinical Pathology 29*, 1014-1015.

Seyfried, P.L., Tobin, R.S., Brown, N.E. and Ness, P.F. (1985a) A prospective study of swimming-related disease. I. Swimming-associated health risk. *American Journal of Public Health 75(9)*, 1068-1070.

Seyfried, P.L., Tobin, R.S., Brown, N.E. and Ness, P.F. (1985b) A prospective study of swimming-related disease. II. Morbidity and the microbiological quality of water. *American Journal of Public Health 75(9)*, 1071-1075.

Simon, G. and Von Graevenitz, A. (1969) Intestinal and water-borne infections due to *Aeromonas hydrophila*. *Public Health Laboratory 27*, 159-161.

Skirrow, M.B. (1986) Communicable disease today - Campylobacter Enteritis. *Environmental Health 94(8)*, 213-214.

Skirrow, M.B. (1987) A demographic survey of campylobacter, salmonella and shigella infections in England. *Epidemiology and Infection 99*, 647-657.

Sorvillo, F.J., Waterman, S.H., Vogt, J.K. and England, B. (1988) Shigellosis associated with recreational water contact in Los Angeles county. *American Journal of Tropical Medicine and Hygiene 38(3)*, 613-617.

Stevenson, A.H. (1953) Studies on bathing water quality and health. *American Journal of Public Health 43*, 529-538.

Sumathipala, R.W. and Morrison, G.W. (1983) Campylobacter enteritis after falling into sewage. *British Medical Journal 286*, 1356.

Symmers, W. (1969) Primary amoebic meningoencephalitis in Britain. *British Medical Journal 4*, 449-454.

Taylor, D.N., McDermott, K.T., Little, J.R., Wells, J.G. and Blaser, M.J. (1983) Campylobacter enteritis from untreated water in the Rocky Mountains. *Annals of Internal Medicine 99*, 38-40.

Turner, P.C., Gammie, A.J., Hollindrake, K. and Codd, G.A. (1990) Pneumonia associated with contact with cyanobacteria. *British Medical Journal 300*, 1440-1441.

Wellings, F.M., Lewis, A.L., Amuso, P.T. and Chang, S.L. (1977) *Naegleria* and water sports. *Lancet i*, 199-200.

Wolfe, M.S. (1979) Managing the patient with Giardiasis: clinical, diagnostic and therapeutic aspects. In: *Waterborne transmission of giardiasis*. EPA-600/9-79-001 pp. 39-52.

Wright, R.A. (1975) Giardial infection from water. *Annals of Internal Medicine 82*, 589-590.

Chapter 7

RECREATIONAL WATERS : A HEALTH RISK?

Rodney Cartwright,
Public Health Laboratory, St Luke's Hospital, Guildford GU1 3NT

INTRODUCTION

Water is and always has been closely associated with both health and disease. It is one of the essentials of life used for drinking and cooking, agriculture and aquaculture, and also provides the medium for many different recreational activities. The transmission of disease through the consumption of water was recognised before knowledge of microorganisms.

Improvements in the health of communities both large and small has been and still is frequently linked to the provision of a safe supply of potable water. The classic observations of John Snow in 1854 culminating in the removal of the Broad Street pump handle to control an outbreak of cholera graphically illustrates the importance of water as a carrier of disease and the importance of epidemiological studies (Snow 1855).

Over the past century careful work has led to the establishment of quality standards for drinking water concerning both microbiological and non microbiological parameters (WHO 1984).

Recreational waters will be an increasing health hazard in the next decade. As more people partake in aquatic sports, many of which involve frequent immersion of the sports person, so the quantifiable hazards will increase. Wind surfing, sailing, canoeing, swimming and many other activities are proving increasingly popular. The hazard to which all are exposed and which will continue to claim lives is that of drowning. In health studies this is an easily studied condition which has a tight definition, does not require complex questionnaires or detailed laboratory studies and for which various preventative measures are internationally agreed. The infection hazards of recreational water has been recognised since the turn of the century and attempts made to quantify the risks.

In 1908, an outbreak of typhoid fever at a Royal Marine Depot at Walmer in Kent (Reece, 1909) was probably caused by bathing in a sewage-polluted swimming pool. The pool in which recruits were given formal instruction in swimming, was filled with sea water though an inflow pipe on an incoming tide, and emptied periodically. It was subject to gross contamination with sewage from two outfalls and the pool water 2-3 days after filling had a bottom sediment containing offensive smelling particles.

The American Joint Committee on Bathing Places (Report, 1949) in the introduction to its 1949 report referred to

> *a growing demand that this committee or some other group of public health workers propose a rigid bacteriological or other standards whereby bathing in certain outdoor bathing waters should be condemned from a public health stand point.*

The Committee argued that no such absolute standard was either practical or desirable. Their advice was not followed and over the subsequent years the United States Environmental Protection Agency (USEPA 1976) adopted a faecal coliform standard for all recreational waters, fresh water and mains. The standard followed studies into the health of bathers in the Ohio River at Dayton, Kentucky (Stevenson 1953) in which there was 32% excess of gastroenteritis among swimmers in water with a medium total coliform density of 2700/100 ml compared with expectation over the whole group of swimmers (Ohio River and a chlorinated swimming pool). Ratio studies of faecal coliforms : total coliforms equated 2700 total to 400 faecal coliforms/100 ml.

This standard was revised in 1986 when the US EPA published new criteria (US EPA 1986) based on the epidemiological studies of Cabelli *et al.* (1979), Cabelli (1983) and Dufour (1984).

These standards are based on an assessment of the health risks and an assessment as to the level of risk which is acceptable.

The European Economic Community Bathing Water Directive (1976) concerning the quality of bathing water defined five microbiological parameters - total coliforms, faecal coliforms, faecal streptococci, salmonella and enteroviruses. There is no information on the derivation of the standards laid down or to the relationship of any between compliance and health risks associated with bathing.

That there is a health risk associated with the recreational use of water is without question. The type of risk, the causative factors and the actions that can be taken to reduce the risk are a matter of debate. As the use of water for recreational purposes increases and as the expectations for risk free activities increases so there is a need to ensure that the relationship between health and the use of recreational waters is viewed objectively.

It is very important that non existent causal relationships are not created. It is easy to make observations which in other circumstances may be cause and effect and to transpose the conclusions, on minimal or no evidence, resulting in unnecessary expenditure. The 'diseases' associated with water must be examined very carefully as many are no more than symptoms developing after immersion in water. It is only by a full and proper evaluation that meaningful actions can be taken to reduce any risks. The improvement of the bacteriological quality of water may be desirable on a number of counts and is always going to be an expensive exercise. Whether the amount of money spent can be justified on cost-benefit terms as far as the health of recreational water users may be a matter of debate. What 'diseases' or symptoms are we trying to prevent? What are the causes? If microbial, where do the microbes originate? How can the human host - parasite balance be altered to the benefit of the host?

Infection results from an upset in the host parasite relationship allowing microorganisms to invade the body. Before assessing the risks associated with recreational waters there are a number of pertinent facts worthy of consideration.

THE HUMAN HOST

Man is a terrestrial animal, well adapted to living in air and for whom, generally speaking, water is a hostile environment. The normal functioning of the human requires that the predominant interface is between the external organs and the air.

The external organs and orifices are developed to function most effectively in an aerial environment. The skin, which in the adult has an area between 1.5 and 2.3 m^2, is a complex biological ecosystem as well as mechanical barrier protecting underlying tissues. It obtains nutrients and moisture from deeper layers and the epithelial surface is continuously being desquamated and replaced. The surface environment is controlled by eccrine sweat glands, sebaceous glands, apocrine glands and hair.

The skin supports an estimated 1012 bacteria as part of the normal microflora. The density varies in different areas of skin, higher counts being obtained in moist areas such as the axilla as compared to the drier forearm (Selwyn and Ellis, 1972). A large increase in the number of bacteria which can be recovered from the skin is observed following bathing or showering falling back to normal in 1-2 hours (Ulrich, 1965). The release of bacteria from the skin is increased by hydration of the stratum corneum - the outer skin layer. Swimmers will release skin bacteria into the water and staphylococci including *Staphylococcus aureus* are consistently present (Robinson and Wood, 1966).

The skin flora is dominated by certain aerobic genera of the family Micrococcaceae, in particular the 'aerobic staphylococci' and *Sarcina* spp. Diptheroids, both lipophilic and non-lipophilic, are next in importance. Gram negative rods including *Mima* spp. and *Alkaligenes* spp. may be residents in moist areas such as the axilla, groins and toe webs. Well recognised pathogens such as *Staph. aureus*, *Pseudomonas aeruginosa* and ß haemolytic streptococci may be part of a permanent or transient flora without being associated with an infection unless there is a disturbance to the anatomical or physiological integrity of the skin.

The skin is frequently considered as an inanimate uniform covering to the body. It is nothing of the sort but rather an organ maintaining a delicate balance between the milieu extérieur and the milieu intérior. Its functionality is largely maintained by the complex arrangement of glands secreting apocrine and eccrine sweat, the stratum corneum and the lipoidal secretions of sebaceous glands which form the majority of the skin surface lipids. The normal flora is influenced by the physiological microclimate the skin surface, a microclimate which can be readily disturbed by external influences. As has been mentioned above, immersion in water even for a short period influences the recovery of bacteria from the skin surface. Such disturbances in the normal balance may well contribute to minor symptoms experienced following recreational activities in water.

The skin covering of most of the body is fortunately very resilient and able to cope with even quite major changes in the external environment, a property much appreciated by regular swimmers.

The eyes are kept moist and healthy by the continuous production of tears from the lacrymal and other glands. Tears act in conjunction with the eyelids to mechanically wash away all particles and bacteria from the conjunctiva. The antimicrobial action is enhanced by the presence of lysozyme in the tears. This mechanism of infection control is very effective unless minor trauma occurs on organisms able to attach to the conjunctiva enter the eye. Chemical irritation may result in local red eye reaction similar to that found in an infection.

The ears are well known for the wax production which provides an important barrier to bacterial infection. Repeated exposure to water results in a loss of the protective wax in the ears making the external canal more likely to be infected by bacteria already present in the canal or introduced in the water giving rise to the condition of otitis externa (Robson and Leung, 1990).

An important aspect of the management of otitis externa is to keep the canal as dry as possible. The excessive use of ear drops even those containing antimicrobial agents usually results in colonisation and even infection due to *Pseudomonas aeruginosa*.

On all body surfaces there is a shifting dynamic boundary between the microroganisms forming a normal resident population and their invasion of the tissues causing an infection. The function of the body of surfaces can also be affected by the external environment which may disturb the normal balance allowing the ingress of microorganisms in the environment. The human body has evolved to function optimally in the gaseous environment of the air. Its function is adversely affected by immersion in water and disease may follow whether or not the water contains microroganisms.

WATER AND THE CHAIN OF INFECTION

Although the development of an infection in an individual is essentially an upset in the host-parasite balance the infecting microbe may spread from one host to another. This feature of communicability provides an extra dimension in the prevention of infection.

The study and control of communicable disease is encapsulated in the chain of infection, the four links of which are the source, the depot or reservoir, the route of spread and the susceptible host. The links need to be elucidated and one or more broken in order to prevent and control the spread of infection. In the consideration of infections associated with recreational waters what is known of the links?

THE SOURCE AND THE DEPOT OR RESERVOIR

These two links merge in the context of recreational water as the prime source for the exposed human is the water. The microbes in the water may be those naturally occurring or those due to contamination by man or animals.

The natural flora of waters consists of microorganisms which have adapted to the environment and are able to maintain steady populations. The species involved will vary according to the type of water. Upland lakes have a very different natural flora from lowland rivers which again will differ from coastal waters. Sewage and agricultural run-off bacteria have a transient life in environmental waters. Concentrations will be higher near the entry sites, reducing due to dilution, the effects of ultraviolet light and the inability to multiply sufficiently to maintain the population. In addition, in areas where there are a large number of people in the water skin organisms will be shed and will add to the microorganism cocktail in the water.

It is not necessary for the organism causing an infection in a recreational water user to originate in the water. The normal flora of the user of recreational water may be the source link. Disturbances in the superficial flora and the outer defence mechanisms following immersion may cause some of these organisms to adopt a pathogenic role. In this scenario the water acts not as a source of organisms but as an aggravating factor. Rapid immersion may result in water entering the nose, upper respiratory tract, eyes and ears carrying superficial organisms as well as those in the water to deeper sites. The chemical content of the water may also cause local disturbances predisposing to infection.

ROUTE OF SPREAD

The interaction between water and the recreational user may be direct contact, inhalation or ingestion. Direct contact with the skin rarely results in a problem although if the skin is abraded or has an open wound local infection may result. Direct contact with the eyes may, however, result in conjunctivitis developing. In the upper respiratory tract water may be forced up the nose especially in activity sports which may be accompanied by forceful immersion.

The spray which may be encountered in recreational activities can be inhaled and carry microorganisms down the respiratory tract.

The ingestion of water is not uncommon during recreational activities. The volume ingested will depend on the circumstances and skill of the individual, but should not approach the amount consumed for drinking purposes.

The three main routes of spread are therefore direct contact, inhalation and ingestion.

SUSCEPTIBLE HOST

Infection will only occur if the exposed subject is unable to cope with the presented microorganisms. The outer defense mechanisms are of prime importance and the integrity of the intact skin is the main defence against skin infections. An abrasion or wound will increase the susceptibility of the host. Prolonged immersion will also alter the effectiveness of the skin barrier. The entry of water into the upper respiratory tract may also increase susceptibility to local infection.

The immune mechanisms provide a deeper level of protection and immunisation programmes aim to reduce host susceptibility. In the context of recreational waters polio vaccination will provide protection against exposure to polio virus in the water. There are no other vaccines which are recommended for users of recreational waters.

WHAT ARE THE RECOGNISED MICROBIOLOGICAL HEALTH HAZARDS ASSOCIATED WITH RECREATIONAL WATERS?

Having indicated that human is a terrestrial animal whose defence mechanisms may be changed by immersion in water and that water may be a source of infection; what microbial diseases are recognised in association with recreational water use?

Although it is important not only to recognise the disease but also the causative organism, much of the published data relates to disease symptoms with a few exceptions. The criteria to be used in assessing causality between environment, exposure and disease were elucidated by Bradford Hill (1965), and are shown in Table 7.1. These criteria should be considered when examining data so that incorrect conclusions are not drawn. The importance of reaching untenable conclusions and the application of inappropriate standards for bathing waters was recognised many years ago (Moore, 1960).

There are a few diseases which can be easily diagnosed and for which the causative agent can be demonstrated. Typhoid fever, (MRC, 1959; Galbraith et al., 1987) shigellosis (Rosenberg et al., 1976), leptospirosis (Waitkins, 1986) and hepatitis A (Bryan et al., 1974) have all been related with bathing in grossly polluted water. Outbreaks of the disease occurred which was then linked to swimming. Norwalk virus infection has been associated with the recreational use of a park lake (Baron et al., 1982) and Adenovirus type 4 pharyngo-conjunctival fever with an inadequately chlorinated swimming pool (D'Angelo et al., 1979). Bathing in natural warm springs has been followed by primary amoebic meningo-encephalitis (Galbraith et al., 1987). There is also unproved but a likely casual relationship for non A non B hepatitis (Ramalingaswami and Purcell, 1988) and cryptosporidiosis (Gallagher et al., 1989). An outbreak of swimmers itch due to cercariae of *Trichobilharzia ocellata* has been described in persons using a Suffolk water sports park. (Eastcott, 1988). Wound infection due to halophilic vibrios has also been recognised. (Armstrong et al., 1983)

In England and Wales, serious and unusual infections are reported to the Communicable Disease Surveillance Centre of the Public Health Laboratory Service and in Scotland to the Communicable Disease Centre (Scotland). Neither centre is

Table 7.1 – Criteria to be used in assessing causality between environmental exposure and disease (Bradford Hill 1965).

Criteria	Explanation
1. Strength of association	Difference in rates of illness between exposed and non-exposed groups. Chi-square test provides a measure.
2. Consistency	Has it been repeatedly observed by different people at different places and times?
3. Specificity of association	A particular type of exposure is linked with a particular site of infection or a particular disease.
4. Temporality	A 'cart and horse' problem - does the exposure predispose to disease or do people susceptible to a particular disease choose that exposure or occupation?.
5. Biological gradient	The more severe the exposure, the greater is the incidence of disease. A dose-response curve can be detected.
6. Plausibility	Does the relationship seem likely in terms of present knowledge? But present knowledge may change.
7. Coherence	The cause and effect interpretation of the data should not conflict with what is know about the biology of the disease.
8. Experiment	Because of an observed association, some action is taken. Is the frequency reduced? This is strong evidence for causation.
9. Analogy	If one agent is shown to cause disease, it would be reasonable to expect it of a related agent.

recording any significant infection associated with the recreational use of water. There is, however, considerable public concern on potential hazards as was shown in the evidence to the House of Commons Environment Committee (1990) in their recent enquiry into the pollution of beaches.

The Medical Research Council report in 1959 on "Sewage Contamination of Bathing Beaches in England and Wales" recognised that, apart from enteric fever and poliomyelitis, minor gastrointestinal illness could occur. The US Public Health Service conducted three studies, two in freshwater and the third in sea water in 1948 and 1950 (Stevenson, 1953). Symptoms relating to eye, ear and respiratory tract exceeded those which were gastrointestinal. Since then there have been many studies in different parts of the world to assess the health hazards to bathers and where possible to relate the incidence of adverse health effects to the bacteriological quality of the water. There has been considerable debate into the epidemiological methods and the bacteriological indicators chosen.

The use of a beach survey with a follow up questionnaire later was pioneered by Cabelli and co-workers (Cabelli, 1983). They were able to demonstrate statistically significant bathing related attacks of gastroenteritis. They also established a relationship between illness and bacterial indicators especially enterococci in the water. More recent studies by Lightfoot (1989) have questioned the validity of the beach survey approach. She considers that a basic design flaw is that the method measures perception not incidence of disease. Fleisher (1991) has reanalysed the data supporting US federal bacteriological water quality criteria governing marine recreational waters and considers that the conclusions of previous analyses are inadequate to support the USEPA's (1986) water quality criteria.

Beach survey type studies have also been undertaken in Egypt (El Sharkawi and Hassan, 1982), Israel (Fattal et al., 1986), Spain (Mujeriego et al., 1982), France (Foulon et al., 1983), Hong Kong (Holmes, 1988, 1989; Cheung et al., 1988, 1990, 1991) and the United Kingdom (Brown et al., 1987; Alexander and Heaven, 1991).

The bacteriological indications which have been principally used are coliforms, faecal coliforms and faecal streptococci. These indicator organisms are those which have for many years been used when assessing the quality of drinking water. It is apposite to consider the section in the first edition in 1934 of United Kingdom Ministry of Health Report No. 71 on 'The Bacteriological Examination of Water Supplies'.

> *Normally the bacteriological examination of water is not directed to the search for specific pathogenic organisms, which are difficult to isolate and usually signify no more than has already appeared by the occurrence of disease among the consumers. The evidence generally sought is (1) an estimate of the number of bacteria of all kinds capable of developing in suitable nutrient media - the greater the number, the greater presumably is the amount of decomposable organic matter present in the water : and (2) the number of bacteria of faecal origin ; the more bacteria, of species inhabiting normally the animal intestine, that are present in the water, the more likely is it that pathogenic intestinal species may gain access to it. The evidence is therefore circumstantial and accordingly, often open to doubt in its interpretation.*

WHAT ARE THEREFORE THE MICROBES WHICH SHOULD BE SOUGHT FOR IN RECREATIONAL WATERS SO THAT DOUBT IN INTERPRETATION CAN BE REDUCED TO A MINIMUM?

There is no easy answer and it may be necessary to continue with the present indicators but recognising their limitations and not ascribing their presence to inappropriate conditions. The indicators should be connected either directly or indirectly to those causing disease.

As the infections acquired through the consumption of contaminated drinking waters are those spread by the faecal oral route it is relevant to use bacteria such as coliforms, faecal coliforms and faecal streptococci which normally reside in the intestinal tract as indicator organisms. They are, in this context, indicators of faecal pollution. Although the gastrointestinal symptoms experienced by bathers may be associated with microorganisms of faecal origin, such an association is not necessarily true for ear, eye and upper respiratory tract symptoms. Seyfried (1985 a,b) in Canada related disease in bathers to the total staphylococcal counts in the water. It is possible that the staphylococcal count in the water may be related to the density of the bathing population, the water acting as the transport system between one bather and another.

At present, there is minimal information on the causes of the common post bathing symptoms. This brings into question the interpretations of studies relating to bacterial counts in water to bathing associated 'disease'.

In spite of these criticisms it is possible to make a few general observations common to the majority of the studies.

1. Bathers especially those immersing their head have a higher incidence of symptomatic illness in the period following exposure.

2. The commonest reported symptoms relate to the eye, ear, upper respiratory tract and gastrointestinal tract.

3. There is no common relationship between water quality and symptoms except for those of the gastrointestinal tract.

4. The illnesses reported are primarily based on symptoms and there are no microbiological studies to identify causative organisms. Indeed there is no evidence that the many symptoms reported have a microbiological cause.

5. The results from individual studies relate to part of the world in which they were undertaken and care should be exercised in applying them to other countries.

In Britain, the Department of the Environment (DoE) has promoted studies to establish the relationships, if any, between microbial quality of coastal waters and the risk to health of bathers.

In view of comments on the beach survey method it was decided to adopt a dual approach. A cohort study would be used in parallel with a modified beach survey. The cohort study would use volunteers who are allocated to bathing and non bathing groups. A medical examination would be undertaken preceding and seven days after exposure. Ear and throat swabs would be taken at the second medical examination and faeces submitted one week and four weeks after exposure. A cohort study was included in order to answer some of the criticisms of the survey approach and would broadly follow the methodology proposed by the WHO (1972:13).

The beach survey would use persons who have themselves chosen to visit a beach and also chosen whether or not to enter the water, all ages could be included and the study would not necessarily have to be on beaches passing the mandatory standards specified in the EEC Bathing Water Directive.

In 1989, a pilot study of both methods was performed at Langland Bay, South Wales and in 1990 a beach survey at Ramsgate, Kent and a cohort study at Moreton Merseyside. The 1991 beach surveys were at Morecambe (Lancaster LA);

Prestatyn (Rhuddlan LA); Lyme Regis (West Dorset LA) and Paignton (Torbay LA). The cohort study will be at Southsea (Portsmouth City). In 1992 four more beach surveys will be conducted and a further cohort study if needed .

The results for 1989 and 1990 showed that the symptoms reported were relatively minor with those relating to the respiratory tract (Pike 1990, 1991), gastrointestinal tract predominating.

In the 1990 beach survey the relative risk for reporting symptoms was increased for those entering the water in the case of one or more symptoms (1.3), gastrointestinal symptoms (1.5) and diarrhoea (1.9). These values altered if water activity sub groups were examined with surfers and divers having more of all symptoms (1.8), respiratory symptoms (2.9) and eye symptoms (2.7) compared to those not entering the water. In the cohort study the relative risk for bathers for flu/cold symptoms was 2.3, chest symptoms 1.8 and gastrointestinal symptoms 1.7 in the seven days after exposure compared to bathers.

Counts of faecal streptococci in the areas of water used by individual bathers in the cohort study were positively associated with increased reporting of ear symptoms, stomach pain, loose motions and gastrointestinal symptoms. Counts of total staphylococci were associated with sore throats, dry coughs, stomach pains and flu/cold among bathers. The intensive bacteriological testing (162 samples plus controls in the 3 hour study period) in the 1990 cohort study demonstrated clearly that bacterial levels could fluctuate between adjacent points at the same time and at a single point at different times. The finding raises the question of single water analysis results. Any study should be accompanied by intensive sampling before conclusions on the microbiological quality of the water are made.

There was no consistent relationship between reporting of symptoms and the results of clinical and laboratory examinations. The relationship of bathing to a health hazard depended on the reporting of perceived symptoms not a clinically observed disease and the studies have not advanced knowledge on the aetiology of minor symptoms.

The microbiological quality of the water in all the DoE studies was assessed using the indicator organisms in the EEC Bathing Water Directive with, in addition, *Pseudomonas aeruginosa*, and total staphylococci.

The quality at Langland Bay and Moreton was within the mandatory values for faecal coliform bacteria and the guidelines for total coliforms and faecal streptococci. Enteroviruses were however detected. At Ramsgate the water quality failed to meet the mandatory values for faecal coliforms, the guideline value for faecal streptococci and the requirements for salmonellae and enteroviruses to be absent.

An unknown variable in both studies is the exact time of immersion in recreational water in the period preceding the study day or initial questionnaire.

When these studies have been completed it should be possible to assess more effectively the health effects associated with the recreational use of coastal waters. Further studies will be necessary for inland waters as it is not possible to apply the results in salt water to fresh water or vice versa.

THE HAZARDS FOR THE NINETIES

It is apparent from all the published studies that the recreational use of water is associated with the development of a variety of symptoms which may or may not be infectious in origin. If there is microbiological evidence of faecal contamination of the water the gastrointestinal symptoms are likely to increase. There is no generally acceptable correlation between either bacterial numbers or species and risk of developing symptoms in spite of standards laid down by various law enforcing agencies. The occurrence of symptoms described will continue for those using recreational water and will depend on host as well as environmental factors.

It is essential that studies into health hazards associated with recreational water should consider both the host as well as the aqueous environment. More information is required on the causative factors of the reported symptoms if sources are to be identified and the first link of the infection chain controlled or broken. It is essential that expensive measures to "improve" the quality of bathing waters are not undertaken until the relationship between the parameters and disease are understood. At the same time it is important that pollutants which could reasonably expect to be associated with disease, if the concentration and exposure is sufficient, should be kept to a minimum or eliminated totally. A conflict of philosophies for those concerned with adding pollutants, especially sewage effluent, to the recreational water environment. Nevertheless naturally occurring organisms such as Leptospira will always remain a hazard.

Recreational waters will always be a health hazard to terrestrial man but in the next decade we should look to understanding in more detail the pathogenesis of the disease so that appropriate control measures can be undertaken.

REFERENCES

Alexander, L.M. and Heaven, A. (1991). *Health risks associated with exposure to seawater contaminated with sewage: the Blackpool Beach Survey 1990.* Environmental Epidemiology Research Unit, Lancaster University, Lancaster LA1 4YB.

Armstrong, C.W., Lake, J.L. and Miller, G.B. (1983). Extraintestinal infections due to halophilic vibrios. *Southern Medical Journal* 76, 571-574.

Baron, R.C., Murphy F.D., Greenberg H.B., Davis C.E., Bregman D.J., Gary G.W., Hughes J.M. and Shonberger (1982). Norwalk gastrointestinal illness. An outbreak associated with swimming in a recreational lake and secondary person to person transmission. *American Journal of Epidemiology 115*, 163-172.

Bradford Hill, A. (1965). The environment and disease: association or causation? Proceedings of the *Royal Society of Medicine 58*, 295-300

Brown, J.M., Cambell, E.A., Rickards A.D. and Wheeler, D. (1987). Sewage pollution of bathing water. *The Lancet ii*, 1208-1209.

Bryan, J.A., Lehmann, J.D., Setiady, I.F. and Hatch, M.H. (1974). An outbreak of Hepatitis A associated with recreational lake water. *American Journal of Epidemiology 99*, 145-153.

Cabelli, V.J. (1983). *Health effects criteria for marine recreational waters. EPA-600/1-80-031*. US Environmental Protection Agency. Health Effects Research Laboratory, Research Triangle Park, North Carolina 27711, 98pp.

Cabelli, V.J., Dufour, A.P., McCabe, L.J. and Levin, M.A. (1983). A marine recreational water quality criterion consistent with indicator concepts and risk analysis. *Journal Water Pollution Control Federation 55*, 1306-1314.

Cheung, W.H.S., Chang, K.C.K. and Hung, R.P.S. (1990). Variations in microbial indicator densities in beach water and health-related assessment of bathing water quality. *Epidemiology and Infection 105*, 139-162.

Council Directive (1976). Council Directive of 8 December 1975 concerning the quality of bathing water (76/160/EEC). *Official Journal of the European Communities 5 February 1976*, No. L31/1.

D'Angelo, L.J., Hierholzer, J.C., Keenlyside, R.A., Anderson, L.J. and Martone, W.J. (1979). Pharyngo-conjunctival fever caused by Adenovirus Type 4: report of a swimming pool - related outbreak with recovery of virus from pool water. *Journal of Infectious Diseases 140*, 42-47.

Dufour, A.P. (1984). *Health effects criteria for fresh recreational waters. EPA 600/1-84-004*. US Environmental Protection Agency, Cincinnati, Ohio 452658.

Eastcott, H.R. (1988). Swimmers' itch, a surfacing problem. An outbreak at a Suffolk water sports park. *PHLS Communicable Disease Report CDR 88/12* (25 March 1988) 3-4 Unpublished.

El Sharkawi, F. and Hassan, M.N.E.R. (1982). The relation between the state of pollution in Alexandria swimming beaches and the occurrence of typhoid among bathers *Bull. High Inst. Pub. Hlth. Alexandria 12*, 337-351.

Fattal, B., Peleg-Olevsky, E., Agurski, T. and Shuval, H.I. (1987). The association between seawater pollution as measured by bacterial indicators and morbidity among bathers at Mediterranean bathing beaches of Israel. *Chemosphere 16 (2/3)*, 556-570.

Foulon, G., Maurin, J., Nguyen Ngoc Quoi and Martin-Boyer, G. (1983). Étude de la morbidité humaine en relation avec la pollution bacteriologique des eaux de bagnade en mer. *Etude préliminarie. Revue Francaise des Sciences de l'Eau 2*, 127-143.

Galbraith, N.S., Barrett, N.J. and Stanwell-Smith, R. (1987). Water and disease after Croydon: a review of waterborne and water associated disease in the United Kingdom 1937-1986. *Journal of the Institution of Water and Environmental Management 1*, 7-21.

Gallagher, M.M., Herndon, J.L., Nims, L.J., Sterling, C.R., Grabowski, D.J. and Hull, H.H. (1989). Cryptosporidiosis and surface waters. *American Journal of Public Health 79*, 39-42.

Holmes, P.R. (1989). Research into health risks at bathing beaches in Hong Kong. *Journal of Water and Environmental Management 3*, 488-495.

House of Commons Environment Committee (1990). *Fourth Report, Pollution of Beaches.* Volumes I, II and III July 1990 HMSO, London.

Lightfoot, N.E. (1989). *A prospective study of swimming related illness at six freshwater beaches in Southern Ontario.* Ph.D. thesis, University of Toronto.

Medical Research Council (1959). *Sewage Contamination of Bathing Beaches in England and Wales.* MRC Memorandum No. 37. HMSO, London.

Moore B (1960). *The risk of infection through bathing in sewage polluted water.* In: Proceedings, 1st International Conference on Waste Disposal in *The Marine Environment*, edited by E.A. Pearson. Pergamon Press, Oxford, p29-38.

Mujeriego, R., Bravo, J.M. and Feliu, M.T. (1982). *Recreation in coastal waters: public health significance.* In Proceedings, IVes Journées Études de Pollutions, Cannes. CIESM Secretariat, Monaco. p585-594.

Pike, E.B. (1990). *Health Effects of sea bathing (ET 9511 SLG). Phase 1 - Pilot Studies at Langland Bay, 1989.* WRc Report DoE 2518-M, Water Research Centre plc, Medmenham 109pp + 2 appendices.

Pike, E.B. (1991). *Health Effects of sea bathing (EM 9511), Phase II - Studies at Ramsgate and Moreton, 1990.* WRc Report DoE 2736-M, Water Research Centre plc, Medmenham 86pp + 3 appendices.

Ramalingaswami, V. and Purcell, R.H. (1988). Waterborne non-A, non-B hepatitis. *Lancet 1.* 571-573

Reece, R.J. (1909) *38th Annual Report to Local Government Board, 1908-9.* Supplement with report of Medical Officer for 1908-9. Appendix A, no. 6, 90.

Report (1934). *The Bacteriological Examination of Water Supplies.* Report No. 71 on Public health and Medical Subjects, Ministry of Health, HMSO, London, 38pp.

Report (1949). *Recommended practice for design, equipment and operation of swimming pools and other public bathing places,* New York : American Public Health Association, 56pp.

Robinson, E.D. and Wood, E.W. (1966). A quantitative and qualitative appraisal of microbial pollution of water by swimmers: a preliminary report. *Journal of Hygiene (Cambridge) 64,* 489-499.

Robson, W.L.M. and Leung, A.K. (1990). Swimming and ear infection. *Journal of the Royal Society of Health 110 (6),* 199-200.

Rosenberg, M.L., Hazlet, K.K., Schaefer, J., Wells, J.G, and Pruneda, R.C. (1976). Shigellosis from swimming. *Journal of the American Medical Association 236,* 1849-1852.

Seyfried, P.L., Tobin, R.S., Brown, N.E. and Ness, P.F. (1985a). A prospective study of swimming-related illness. I. Swimming associated health risk. *American Journal of Public Health 75,* 1068-1070.

Seyfried, P.L., Tobin, R.S., Brown, N.E. and Ness, P.F. (1985b). A prospective study of swimming-related illness. II. Morbidity and the microbiological quality of water.. *American Journal of Public Health 75,* 1071-1075.

Snow, J. (1855) *On the mode of communication of cholera.* Churchill, London.

Stevenson, A.H. (1953). Studies on bathing water quality and health. *American Journal of Public Health 43,* 529-538.

Ulrich, J.A. (1965). *Dynamics of bacterial skin populations.* In Maibach, H.I. and Hildick-Smith, G. (Eds.) *Skin Bacteria and their Role in Infections.* New York, McGraw-Hill.

USEPA (1976). *Quality Criteria for Water.* US Environmental Protection Agency, Washington DC.

USEPA (1986). *Ambient Water Quality for Bacteria. EPA 440/5-84-002,* US Environmental Protection Agency, Washington DC, 18pp.

Waitkins, S.A. (1986). Leptospirosis in man. British Isles : 1984. *British Medical Journal* 292, (17 May 1986), 1324.

WHO (1972). *Health criteria for the quality of recreational waters with special reference to coastal waters and beaches.* Ostend Belgium 13-17 March, 26pp.

WHO (1984). *Guidelines for Drinking Water Quality.* World Health Organisation, Geneva.

Chapter 8

STATISTICAL ASPECTS OF MICROBIAL POPULATIONS IN RECREATIONAL WATERS

Edmund B Pike
Water Research Centre, Medmenham, UK

INTRODUCTION

Microbiological analyses lack the precision of chemical analyses and of physical measurements. Although extremely sensitive and specific, they are inaccurate, because the true count cannot be obtained. Methods which rely on growth responses, such as colony counts or multiple-tube dilution counts, detect only those bacteria which are able to grow under the conditions provided. A growth response may be given by a clump of bacteria, as well as by a single cell. These difficulties should be regarded as a challenge by the microbiologist and not as an excuse for poor technique, under the justification that, because the methods are so imprecise, a little more will not make much difference. Analyses have a customer, who will need to be satisfied by the reliability of the analyses, since the results will be used for making decisions.

This paper reflects the Author's interest over 25 years in studying microbial pollution of surface waters and in the microbial ecology of sewage treatment and when he was responsible for teams of microbiologists working in field laboratories. It will consider the purposes for which microbiological analyses are carried out and consider natural sources of variability in microbial numbers in water and sources of error in sampling and in the laboratory. It suggests ways in which natural variability and analytical errors can be assessed.

Statistics can be defined as the precise study of imprecision. Analysis of variance (ANOVA) provides a way of assessing independent sources of variability in factorially designed trials. Finally, an approach to analytical quality control (AQC) of microbiological analyses is suggested.

THE PURPOSES OF SAMPLING AND ANALYSIS

All analyses have a purpose. It is important that the purposes and the uses to which the results will be put should be discussed with the client beforehand, so that effective and economical sampling strategies can be worked out. Typical reasons for requesting analyses are:

(a) Site surveys. Assessment of sources of faecal contamination from human and animal sources, prior to development of a facility, such as for recreation or for supply of drinking water.
(b) Assessing compliance with a water quality standard or guideline.
(c) Assessing the need for remedial action after failure to meet a standard, or where water is implicated in an outbreak of infectious disease.

These purposes will need differing choices of determinand and of sampling strategy. With (b), the determinand will usually be specified, if not the method of analysis. With (a) and (c), an informed decision will be needed. Detection of enteric pathogens in water is relatively costly and difficult, compared with enumeration of faecal indicator bacteria and, except in the case of well defined outbreaks of water-borne disease, pathogens may occur sporadically or not at all. In the case of (a) and (c), inspection on foot to assess sources of pollution is as important as microbiological analysis. The frequency and times of sampling should reflect the times of day, the season and the periods during which the facilities are used.

Single determinations are of very limited value, since estimates of central tendency and variability are needed to define prevailing conditions at a site and to judge compliance. Counts of micro-organisms taken from a region, or even from a single sampling point, are usually distributed approximately log-normally, i.e., the logarithms of count are normally distributed (Gameson et al., 1970). This means that the correct descriptor of central tendency is the geometric mean, for which the median (the 50th percentile) can be substituted, when very high or low values are indeterminate. Arithmetic averages of counts give falsely high values of central tendency. Likewise, variability is best expressed as the standard deviation of log counts. Standard deviations of untransformed counts are excessively high, numerically, and dependent upon the mean. Ranges of counts are of limited value, since they are influenced by the number of data points as well as upon variability.

Log-normality implies that statistical analysis should be carried out upon logarithms of data points, to avoid false conclusions being drawn.

The variability of bacterial counts at a sampling station will depend upon the size and location of polluting discharges, upon weather, season and the duration of the period of measurement and upon the degree of hydraulic mixing. In the author's experience, values of about 0.7 are usual for the $\log_{10}$ standard deviation of coliform counts in coastal, estuarine and lake waters, with somewhat lower values (0.6) for rivers. Broadly similar values were found by Gameson (1978) for coastal and river waters. The U.S. Environmental Protection Agency's (1986) criteria for recreational waters used for bathing give typical values of 0.7 for marine waters and 0.4 for freshwater beaches.

It should be noted that the variability observed in any set of data is the sum of individual components caused by factors operating independently. Variance (the square of standard deviation) is additive and ANOVA provides a method for breaking the total variance down into separate components.

FACTORS RESPONSIBLE FOR VARIABILITY IN WATER QUALITY

At any sampling point, the true water quality will be influenced in the following ways:

- Periodically : seasonal, tidal and diurnal effects, including patterns of sewage discharge
- Spatially : e.g. at the water's edge, offshore or at depths, where different hydraulic regimes operate.
- Irregularly : e.g. because of storms, irregular discharges or effects of wind.

Polluting micro-organisms, released into water, are subjected to dilution, dispersion, sedimentation and mortality. Mixing is not instantaneous or complete. Sedimentation only affects those micro-organisms which are adsorbed to particles; adsorption is strongest to clay or silt particles and is least to sand. Mortality will occur in darkness, but is most rapid in daylight, being brought about by the longer wavelengths of ultra-violet radiation in sunlight (300-400nm). Even in clear water, UV-induced mortality occurs only in the first few metres depth of water. Increases in salinity and temperature increase the rate of mortality. The presence of nutrients in water may retard mortality, or, in extreme circumstances, permit growth.

The log-normal distribution is common in biological measurements. It is interesting to consider that those mechanisms which can generate it exactly are found in natural waters. These are the application of first-order rate processes (e.g., diffusion, mortality or growth, light extinction) to an underlying normally distributed population, or repeated fluctuations in such processes applied to all members of the microbial population (Koch, 1966).

SOURCES OF ERROR IN THE LABORATORY

Estimates of microbial counts can be affected critically during the following operations:

- Sampling - method and position of sampling.
- Transport and storage of samples before analysis.
- Dilution of samples before analysis.

Incubation conditions - resuscitation, culture medium, temperature.

Counting of colonies.

The position of sampling should be standardised and may be specified in regulations or guidelines. Surface scums may carry a higher density of bacteria than waters underneath, although this is not necessarily so. It should be obligatory for samples to be placed in a light-tight container immediately after taking and for them to be kept in darkness until analysed. Exposure of bottles to direct sunlight can reduce counts ten-fold in about 20-30 minutes. There is no safe period of storage before analysis, since counts will change, either decreasing or increasing during storage. Changes can be minimised by placing bottles in contact with melting ice, so that they are rapidly cooled. Sea water samples taken in June - October at Sidmouth, Devon, declined by a factor of 0.70 during storage in the dark for 6 hours at ambient temperature, equivalent to a decimal reduction time (T90) of 39 hours (Gameson *et al.*, 1967).

Selective and enrichment media used for isolating faecal bacteria necessarily contain selective inhibitors of growth, which enable the required bacteria to outgrow others, which are suppressed. All faecal bacteria in natural waters are exposed to environmental stress (e.g., lack of nutrients, metabolic damage from sunlight or salinity) and may be suppressed to some extent during incubation. The concentration of selective agent (e.g., bile salt, sodium lauryl sulphate or sodium azide) is critical. Over-sterilization of culture media by heat may either reduce the selective action, or generate inhibitory agents. Selectivity is also determined by incubation temperature, particularly in the determination of thermo-tolerant coliform bacteria or faecal streptococci at 44°C. The recovery of environmentally stressed bacteria is assisted by a compulsory period of resuscitation at 30°C for 4 hours, before incubation at 44°C. Use of 0.1 or 1 per cent peptone water as a diluent before membrane filtration, will also improve recovery. If samples are filtered undiluted, the effect can be obtained by drawing through 10 ml of peptone water, after filtering.

Particular attention must be given to the operation of dual-temperature, time-cycling incubators, used for providing resuscitation at 30°C. Temperature recording is desirable, but the recorder must be accurate and immediate action must be taken to correct errant temperatures. The author recalls one incident in which apparently perfect temperature control was '*caused*' by a thermometer with a broken bulb and another in which the standard thermometer, used for checking, was greatly in error.

If bacteria are randomly dispersed in water, then their numbers in samples of water taken for analysis and the resulting counts of colonies on plates or membranes should be distributed Poissonianly, with variance equal to the mean. If so, then the precision will increase as the colony count increases. If precision is described as the coefficient of variation (standard deviation as a percentage of the mean) then it is inversely proportional to the square root of the mean, for a Poisson distribution. Quadrupling the colonies counted halves the coefficient of variation. In cases where descending dilutions or different volumes of sample are plated or filtered, a useful increase in precision can be obtained by adding the counts of colonies on all countable plates and dividing by the total volume examined.

In practice, results of replicate determinations or considerations of the ratios between counts obtained from different volumes of water, show that there are deviations from Poissonian distribution (Gameson 1983, Tillett and Farrington 1991). Without special experimentation, it is only possible to speculate upon the causes, which may include the following:

> Aggregation of bacteria in the original sample.
>
> Adhesion of bacteria to surfaces during pipetting and handling.
> For low counts of colonies - absence of synergistic bacteria.
>
> For very high colony counts, crowding and suppression.
>
> Miscounting and visual fatigue.

In Fig.8.1, the author plated varying and replicated volumes of activated sludge. The colony counts show decreasing variability as they increase, but not to the extent which would be predicted by Poissonian theory until, with counts exceeding 1000 colonies, partial suppression appears (DOE, 1972). Sub-sampling or dilution of

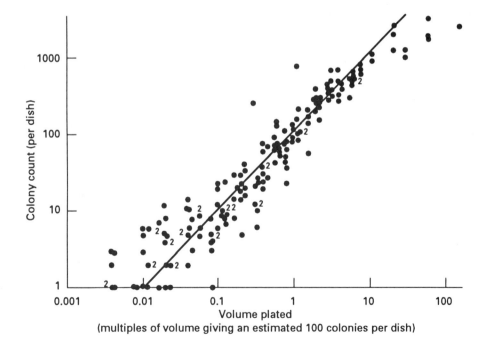

Figure 8.1 – Relationship between counts of colonies and volumes of activated sludge mixed liquor plated by spreading on casitone-glycerol-yeast extract agar; incubation at 22°C for 6 days (DoE 1972) .The line of equality and numbers of overlapping points are shown.

original samples introduces additional loss of precision, as expected from Poissonian theory (Tillett and Farrington, 1991), as does estimating counts from sectors of overcrowded plates.

The author once worked in a team with two colour-blind scientists. One never counted colonies, when colour differentiation was required and meticulously labelled and arranged bottles of coloured media and recently poured plates. The other, a temporary worker, was found confusing peptone water (straw colour) with Teepol broth (red) and then admitted his problem. A commoner problem with colour differentiation occurs with crowded plates and the conjunction of acid- and alkali-producing colonies.

MEASUREMENTS OF ERROR IN ANALYSIS

Proper instruction, supervision and working practices should minimise analytical errors. However, application of analytical quality assurance makes it desirable to be able to measure analytical errors within replicate determinations, between different staff and between laboratories. The irreducible sources of variability are those caused by environmental factors affecting the water itself and those largely Poissonian, inherent in sampling, sub-sampling and in plating bacteria distributed in the water. The remaining variability is reducible by strict control over the methods of analysis. These sources of variability can be measured by factorial design of replicated analyses and by applying ANOVA to the data. A trial involving tenfold replication of membrane filtration for total coliform bacteria in ten samples, taken on each of ten dates at a marine beach, gave a standard deviation of 0.1073 as the residual replication error of log counts (Fleisher and McFadden, 1980). Nested ANOVA showed that 95% of the total variance was attributable to sampling dates.

This example shows that replication of analyses can be used to estimate residual variance within samples, once the variances attributable to such factors as different sampling stations, or times of day have been extracted by ANOVA. A simple approach to measuring within samples variability in the laboratory would be to request replicate analyses of the first and the last samples handled each day and to use the individual colony counts to calculate the residual, within samples variance for comparative purposes.

An example of the use of ANOVA is given in Table 8.1. A more complex example could include samples taken on different days at different sites. Such examples show that small, but probably significant differences in water quality can be missed if the analytical techniques are poor. It is the author's experience that significant interactions are often revealed between the factors of times of day, dates and locations, doubtlessly because dispersion of faecal bacteria is affected by tidal currents as well as by diurnal factors and position onshore.

Table 8.1 – ANOVA for triplicate counts of thermo-tolerant coliform bacteria at two beach sites, 40m apart, sampled at 20-minute intervals on seven occasions.

Factors	Degrees of freedom	Sum of squares	Mean squares	F square
Runs	6	4.015491	0.669249	8.38**
Locations	1	0.110264	0.110264	1.38
Runs x Locations	6	0.699236	0.116539	1.46
Residual	28	2.236976	0.079872	
Total	41	7.061967		

** Significant at P < 0.01
Standard deviation of residual (between replicates) = 0.283

THE APPROACH TO MICROBIOLOGICAL AQC

The same broad approaches used for AQC of chemical analyses (Cheeseman and Wilson, 1989) can be adapted for microbiological analysis. The procedures to be followed can be summarised thus:

> Establish working group.
>
> Define determinand, limit, accuracy required.
>
> Choose or develop methods with low bias and adequate precision
>
> Specify methods completely and unambiguously.
>
> Estimate inter-laboratory precision and improve to meet target.
>
> Establish reliable standard inocula.
>
> Set up quality control charts.
>
> Carry out inter-laboratory trials and improve agreement to meet target.

It may be thought difficult to implement certain of these stages. However, the statistical approaches discussed above should enable *adequate precision* to be defined and they should provide a way for measuring improvements within and between operators and laboratories. Standard inocula are, or will be, available in two schemes. The Public Health Laboratory Service has a scheme for distributing standard water samples (from the Public Health Laboratory, Newcastle-upon-Tyne), in which the position and distribution of results is notified to participating laboratories. In the scheme being developed by RIVM (Bilthoven, The Netherlands), bacterial strains of certified density, preserved in gelatin capsules of skim-milk powder, will be provided to laboratories.

As a final note, the precision, accuracy and reliability of microbiological analysis, lie in good training, supervision and in the application of standardised techniques, from sampling on site, to the calculation and presentation of results. Good results in inter-laboratory comparability exercises and certification can only result from continuous attention to these points.

REFERENCES

Cheeseman, R.V. and Wilson, A.L. (1989) *A manual on analytical quality control for the water industry.* Report No. NS 30, Water Research Centre, Medmenham.

DoE (1972) *Water Pollution Research 1971.* Department of the Environment. HMSO, London, pp.62-63.

Fleisher, J.M. and McFadden, R.T. (1980) Obtaining precise estimates in coliform enumeration. *Water Research 14*, 477-483.

Gameson, A.L.H.(1979) *Investigations of sewage discharges to some British coastal waters. Chapter 5. Bacterial distributions, Part 1.* Technical Report 79, Water Research Centre, Medmenham.

Gameson, A.L.H. (1983) *Investigations of sewage discharges to some British coastal waters. Chapter 3. Bacteriological enumeration procedures, Part 2.* Technical Report 193, Water Research Centre, Medmenham.

Gameson, A.L.H., Bufton, A.W.J. and Gould, D.J. (1967) Studies of the coastal distribution of bacteria in the vicinity of a sea outfall. *Water Pollution Control 66*, 501-523.

Gameson, A.L.H., Munro, D. and Pike, E.B. (1970) *Effects of certain parameters on bacterial pollution at a coastal site.* In Symposium, Water Pollution Control in Coastal Areas, Bournemouth, 19-21 May, 1970. Institute of Water Pollution Control, Maidstone, Paper 4.

Koch, A.H. (1966) The logarithm in biology. Mechanisms generating the log-normal distribution exactly. *Journal of Theoretical Biology 12*, 276- 290.

Tillett, H.E. and Farrington, C.P. (1991) Inaccuracy of counts of bacteria in water or other samples: effects of pre-dilution. *Letters in Applied Microbiology 13*, 168-170.

US Environmental Protection Agency (1986) *Ambient Water Quality Criteria For Bacteria. EPA 440/5-84-002.* Office of Water Regulation and Standards, US Environmental Protection Agency, Washington DC.

Chapter 9

US FEDERAL BACTERIOLOGICAL WATER QUALITY STANDARDS : A RE-ANALYSIS

Jay M Fleisher
State University of New York, Brooklyn, NY and
Senior Research Fellow, CREH, University of Wales, Lampeter

INTRODUCTION

Current federal bacteriological water quality criteria governing the sanitary quality of marine recreational waters were directly derived from the results of a large prospective epidemiological study conducted for the United States Environmental Protection Agency (EPA) by Cabelli *et al.* (1982). This study, hereafter referred to as the EPA study, derived a mathematical relationship between enterococci densities and gastroenteritis among swimmers. Unfortunately, this study contained serious methodological weaknesses. It is the purpose of this paper to explore these methodological weaknesses and to present a re-analysis of the data upon which current federal bacteriological criteria governing marine recreational waters are based. This re-analysis will show that the mathematical relationship between enterococci density and gastroenteritis reported in the EPA study is of questionable validity so that current federal bacteriological water quality criteria governing marine recreational waters are not based upon strong enough evidence to support their continued use.

THE EPA STUDY DESIGN

The EPA study was conducted on selected weekends over a 6-year period (1973 to 1978) and included 25,442 study subjects. Two marine bathing areas (New York City, NY and Boston, Massachusetts) and one *brackish* area (Lake Pontchartrain, Louisiana) were chosen as study sites. Indicator organism densities were sampled on each trial day. The exposed group consisted of bathers who completely immersed their heads in the water (swimmers). The reference group included bathers who either did not enter the water or who entered the water but did not immerse their heads (non-swimmers). Exposure status was determined by interview conducted on each trial day. Incidence of gastroenteritis was obtained via telephone interview 8 to 10 days after the initial interview. Gastrointestinal (GI) ailments were classified into 2 types: Total GI symptoms and Highly Credible GI symptoms. Total GI symptoms included all incidences of vomiting, diarrhoea, stomachache, and nausea reported at telephone interview. Highly Credible GI symptoms include all cases of vomiting, instances of diarrhoea accompanied by a fever or that were disabling, or cases of nausea or stomach ache accompanied by a fever. Individuals who swam in the midweek prior to or after the weekend under study were excluded from the analysis. The method of analysis used was least squares linear regression and consisted of regressing the Log_{10} enterococci density on the incidence rate difference of gastroenteritis among swimmers vs non-swimmers. Rather than use each trial date as the basic unit of measurement, trial dates were clustered according to *natural breaks in measured indicator organism densities* yielding only 18 data points upon which the regression analysis was based.

METHODOLOGICAL WEAKNESSES IN THE EPA STUDY

A major methodological weakness incorporated within the EPA study stems from the fact that the results obtained for both marine and *brackish* water locations were pooled in the final analysis of the data. At the time the study was conducted there was some previously published evidence of an association between indicator organism density and illness among swimmers in fresh waters but no evidence of such an association in marine waters (Stevenson, 1953; PHLS, 1953). Subsequent epidemiological studies have shown that associations between indicator organism density and illness among swimmers vary markedly between fresh and marine water locations (USEPA, 1983; Seyfried *et al.*, 1985; Ferley *et al.*, 1989). The *brackish* water location used in the EPA study (Lake Pontchartrain) more closely approximated a fresh water location. This is based on the fact that during the study period salinity at this location ranged from 1.1 to 4.8 parts per thousand (US Army Corp of Engineers, pers. comm., 1989). Assuming the salinity of sea water to be 35 parts per thousand, this location would, for all practical purposes, be considered a fresh bathing water location (Velz, 1970). As will be shown later in this report, failure to control for fresh vs marine water locations in the analysis used in the EPA study cast serious doubt on the reported findings.

A related issue deals with the possibility that the *true* mathematical relationship between indicator organism density and illness among swimmers will vary from location to location within both fresh and marine bathing waters. Factors such as local differences in resistance or immunity to pathogenic micro-organisms among swimmers, differences in the composition of the underlying pathogens, and differences in the relationship between indicator organism density and underlying pathogen density lend biological support for possible differences in mathematical relationships between enterococci density and gastroenteritis at different marine water locations within the United States. If such differences are shown to be significant, the appropriateness of uniform federal recreational water quality criteria becomes questionable. The analysis used in the EPA study did not test, or control for, such local variation. As will later be shown, significant variation in the mathematical relationship between enterococci density and gastroenteritis among swimmers did indeed exist among the three EPA study locations.

The choice of using absolute vs relative measures of effect to express the results of epidemiological studies is frequently a difficult one. The authors of the EPA study no doubt chose to measure the absolute effect (rate difference) because of its frequent use in public health planning. There are several reasons why use of the rate difference as the measure of effect seems inappropriate in the context of the EPA study's objectives. The first stems from the fact that the rate difference is not affected by changes in the baseline incidence of disease. The rate ratio, however, is dependent on the magnitude of the baseline incidence rate and, as such, controls for changes in the rates of illness among the unexposed. Table 9.1 shows the data upon which the EPA study's analysis is based. Inspection of Table 9.1 shows that the rates of gastroenteritis among the unexposed (non-swimmers) varied greatly. Use of the rate differences of gastroenteritis (exposed-unexposed) as the dependent variable in the EPA study's regression analysis will not adjust for possible effects caused by changing incidence of gastroenteritis in the unexposed relative to observed rate differences at similar enterococci densities and therefore is not the most appropriate method of analysis.

A second reason why use of the rate difference was inappropriate in the context of the EPA study's objectives stems from the fact that the use of an absolute measure of effect is only valid to the extent to which the observed difference in rates are truly attributable to the exposure (Rothman, 1986). Reporting the results of analysis that used the rate difference as the measure of effect would, by definition, attribute all excess risk exclusively to the exposure. Since the mathematical relationship reported in the EPA study had little support in the existing epidemiological literature, it would be imprudent to ascribe any observed difference in the rates of gastroenteritis among swimmers vs non-swimmers solely to the exposure (enterococci density) without assessment for the presence of residual confounding factors. Indeed, the authors of the EPA study failed properly to interpret the results of their regression analysis. The R^2 value of the regression equation derived in the EPA study equalled 0.56. Although this indicated that only 56 percent of the variation in the rate differences of gastroenteritis could be explained by changing enterococci density, this regression equation was used to derive the maximum allowable enterococci densities contained in current federal criteria without any adjustment relative to the observed R^2 value. This could result in an over-estimate of risk and is therefore inappropriate.

A third argument for use of a relative measure of effect deals with the statistical stability of the incidence rates estimated by the EPA study. As previously mentioned, the authors of the EPA study combined individual trial dates. The reason for this clustering was that some of the individual trial dates did not have enough non-swimmers to calculate statistically stable rates of gastroenteritis among non-swimmers. Such clustering of trial dates inevitably led to a loss of information. Had relative risk been used as the measure of effect, an analysis of the data could have been conducted (e.g., using logistic regression analysis) that would have made such clustering of trial dates unnecessary. This is especially important since the clustering of trial dates resulted in the fact that current federal bacteriological water quality criteria governing marine recreational waters were directly derived from a least squares linear regression analysis of only 18 data points. Lastly, it should be noted that in the medico-legal or regulatory arena, relative risk is a more useful measure of effect than absolute risk (Hoffman, 1984). Unlike measures of absolute risk whose inference is to the population, statistical procedures that model associations in terms of relative risk (e.g. Logistic Regression Analysis) can be used to focus on the individual's probability of developing disease. In this regard, it should be noted that in the two epidemiological studies finding associations between enterococci density and swimming-associated illness published subsequent to the EPA study, both presented their results in terms of relative risk (Seyfried *et al.*, 1985; Ferley *et al.*, 1989).

DATA ANALYSIS WEAKNESSES IN THE EPA STUDY

The actual analysis conducted by the authors of the EPA study also contained serious flaws. As previously mentioned, the method of analysis was to regress the mean enterococci density from each of the 18 clustered trial dates on the corresponding rate differences of gastroenteritis among swimmers vs non-swimmers. An unweighted least squares linear regression procedure was used even though a weighted regression procedure would have been more appropriate (Kleinbaum *et al.*, 1988). Indicator organism density was Log_{10} transformed prior to analysis yielding the following analytical model:

$$Y = a + b (Log_{10} X)$$

Where X = Mean enterococci density 100ml^{-1}
 Y = Rate difference of GI symptoms 1000 persons^{-1}
 (swimmers - non-swimmers)

This analysis was done separately for *Total* and *Highly Credible* GI symptoms. As previously discussed, the actual data used comprised 18 clustered trial dates as shown in Table 9.1.

Figure 9.1 shows the results of these regression analyses. Inspection of Figure 9.1 shows significant regression coefficients for both Total and Highly Credible GI symptoms. The R^2 values were 0.67 and 0.56 for Total and Highly Credible GI symptoms respectively. These values indicate a moderately good fit of the data to the regression line but the fact that Total GI symptoms fit somewhat better is troubling. If indeed Highly Credible symptoms were a more reliable measure of swimming-associated illness than Total GI symptoms, one would expect a better fit in the Highly Credible category. Figure 9.2 shows a regression of enterococci density on the rate ratios (swimmers/non-swimmers) derived from Table 9.1 for each

Table 9.1 – Data used in regression analysis conducted by the EPA study[a].

Summary of the mean enterococcus density-gastrointestinal (GI) symptom rate relationships among swimmers, non-swimmers and residuals obtained from clustered trials in studies on swimming-associated gastrointestinal illness, 1973—1978

Study	Beach	Year	Enterococcus density/100ml Mean	Range	Trial (days) clustered	S	NS	Total GI S	NS	R	Highly credible GI S	NS	R
New York City	Rockaways	1973†	21.8	1.2-59	8	484	197	81	46	35	30.4	15.2	15.2
	Coney Island		91.2	6-186	8	474	167	72	24	48*	46.4	18.0	28.4
		1974	3.6	2-5	3	1391	711	27	23	4	7.6	4.2	3.4
			7.0	7	3	951	1009	38	34	4	10.5	6.9	3.6
			13.5	10-17	2	625	419	42	17	25*	16.0	2.4	13.6
			31.5	30-33	2	831	440	43	23	20	18.1		18.1"
		1975	5.7	2-11	14	2232	935	63	55	8	18.8	19.3	-0.5
			20.3	14-38	10	1896	678	59	37	22*	14.8	7.4	7.4
			154	86-298	4	579	191	60	31	29	34.5		34.5
Lake Pont-chartrain, LA	Levee	1977	44	9.7-88	8	874	451	86	51	35*	32.0	11.1	20.9*
			224	190-249	4	720	456	108	50	58**	31.9	8.8	23.1*
			495	344-711	2	895	464	108	54	54**	35.8	8.6	27.2**
	Levee	1978	11.1	3-30	8	1230	415	75	34	41**	36.6	14.5	22.1*
	Fontainbleau		14.4	3-33	5	248	303	81	63	18	44.3	23.1	21.2
	Levee		142	67-303	4	801	322	112	50	62**	42.4	15.5	26.9*
Boston, MA	Revere	1978	4.3	2-6	3	697	529	83	66	17	23.0	11.0	12.0
	Nahant		7.3	6-9	2	1130	1099	71	67	4	33.0	28.0	5.0
	Revere		12.0	12	1	222	376	108	74	34*	41.0	13.0	28.0*

S : Swimmers   NS : Non-swimmers   R : Residuals
*$p < 0.05$;   **$p < 0.01$.
†Study population too small to cluster trials by similar indicator densities.
[a] From Cabelli et al. (1982), Table 5, Page 611.

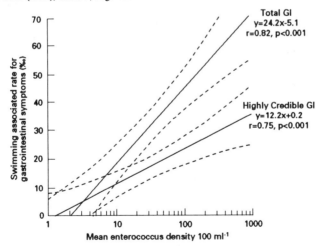

Figure 9.1 – Regression lines depicting the association between the rate difference of GI symptons among swimmers and non-swimmers and enterococci density as reported by the EPA study.
From Cabelli et al. (1982), Figure 2, Page 614.

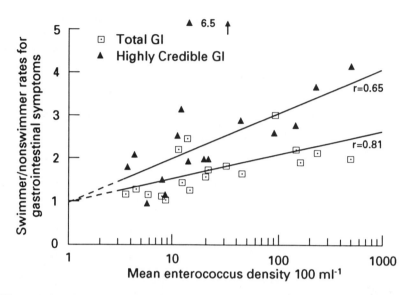

Figure 9.2 – Regression lines depicting the association between the rate ratio of GI symptoms (swimmers/non-swimmers) and enterococci densities as reported by the EPA study.
From Cabelli et al. (1982), Figure 3, Page 615.

of the 18 data points for both Total and Highly Credible GI symptoms (This analysis was reported in the EPA study but not incorporated in the adopted water quality criteria). Table 9.2 shows the actual regression analysis calculated from the data in Table 9.1 by this author for the regression lines depicted in Figure 9.2. These calculations were not shown in the EPA study. It is of considerable interest that 3 data points were arbitrarily dropped by the authors of the EPA study in the regression analysis for Highly Credible GI symptoms when using the rate ratio as the dependent variable but were included in the regression analysis that used the rate difference as the dependent variable. Two of the three data points that were omitted corresponded to trial clusters that had no reported GI symptoms among non-swimmers. (The third was omitted due to an unusually low non-swimmer rate). An alternative to exclusion would be to use the average rate of GI symptoms reported among non-swimmers for the year and location of these two *missing* data points. This method would yield expected non-swimmer rates of 4.5 and 13.3 per thousand for these two omitted data points. Table 9.3 shows the results of a regression of the Log_{10} enterococci density on the rate ratio (swimmer/non-swimmer) using the full data set incorporating the above two estimates of non-swimmer GI rates in lieu of exclusion. Inspection of Table 9.3 shows that, when the data for Highly Credible symptoms are re-analyzed in this manner, the regression coefficient changed considerably and the equation is no longer significant ($p>0.05$). Although it can be argued that the methods used to derive the analyses shown in Table 9.3 are also arbitrary, the striking differences

Table 9.2 – Linear regression of the $\log_{10}$ enterococci density on the rate ratios for total and highly credible GI symptoms. Analysis corresponds to the EPA's original analysis (see text for further details).

Total GI symptons

Variable	Regression coefficient	R^2
$\log_{10}$ Enterococci Density	0.5507	0.42
Intercept	0.9747	

Analysis of variance

	SS	df	MS	F	P
Regression	2.0716	1	2.0716	1.50	0.004
Residual	2.8815	16	0.18G1		

Highly Credible GI symptons

Variable	Regression coefficient	R^2
$\log_{10}$ Enterococci Density	1.0766	0.65
Intercept	0.8806	

Analysis of variance

	SS	df	MS	F	P
Regression	7.0729	1	7.0729	24.57	0.0003
Residual	3.7417	13	0.2878		

Table 9.3 – Linear regression of the $\log_{10}$ enterococci density on the rate ratios for highly credible GI symptoms. Regression shown below is a re-analysis of the EPA's original analysis without the exclusion of 3 data points (see text for further explanation of this analysis).

Variable	Regression coefficient	R^2
$\log_{10}$ Enterococci Density	0.8398	0.16
Intercept	1.5131	

Analysis of variance

	SS	df	MS	F	P
Regression	4.8170	1	4.8170	3.030.	0.10
Residual	25.4038	16	1.5877		

between this analysis and that reported by the EPA study highlight the enormous effect that can be caused by minor manipulation of the data. This phenomena could have considerable relevance to the outcome reported by the EPA study; specifically, the potential effect of clustering sample dates prior to analysis.

Using the regression lines shown in Figures 9.1 and 9.2, the EPA study concluded that swimming in water containing as little as 10 enterococci/100 ml of sample represented an absolute risk of GI illness of 10/1,000 swimmers and a relative risk of 2.0, and that swimming in even marginally polluted marine waters represents a significant route of transmission of gastroenteritis. Moreover, current U.S. EPA bacteriological criteria governing the sanitary quality of marine waters were computed directly from the regression equation for Highly Credible symptoms shown in Figure 9.1.

A RE-ANALYSIS

Using the data for highly credible GI symptoms shown in Table 9.1, unconditional multiple logistic regression analysis (Breslow and Day, 1980) was used to ascertain and quantify a possible association between gastroenteritis among swimmers and enterococci density. Since animal infectivity studies conducted with most infectious agents yield sigmoid dose response curves (USEPA, 1983), use of the logistic model has biological support. The rates of Highly Credible GI symptoms for both swimmers and non-swimmers shown in Table 9.1 were converted to the actual number of swimmers or non-swimmers experiencing a Highly Credible GI symptom on each of the 18 clustered trial dates as well as the number not experiencing such symptoms. This was accomplished by multiplying the illness rate reported for swimmers or non-swimmers by the number of swimmers or non-swimmers at risk on a particular clustered trial date. In this manner logistic regression analysis could be used to evaluate the relationship between exposure and outcome variables.

The Logistic Regression Model specifies that the probability of an individual contacting a disease depends on a set of variables $X_1, X_2, ..., X_k$ such that:

$$\text{Probability of Disease} = \frac{1}{1 + e^{-(B_0 + B_1 X_1 + \cdots + B_k X_k)}}$$

Where;

$$B_0 + B_1 X_1 + \ldots + B_K X_K = \text{Log Odds of Disease}$$

The regression coefficients (B) are derived using methods of maximum likelihood. Logistic Regression Analysis is one of the most widely used statistical models in epidemiological studies seeking to quantify exposure-disease associations (Breslow and Day, 1980; BMDP, 1988).

In order to simplify model building all logistic modeling was restricted to swimmers only. The two main effects used in the logistic regression model were enterococci density and study location (New York, Boston and Lake Pontchartrain).

The logistic regression models were computed using the step-wise procedure contained in the BMDP package of statistical software (BMDP, 1988). Model building, however, was conducted in a non-stepwise fashion by controlling the sequence of independent variables entering the analysis. In this manner, the stepwise procedure was over-ridden allowing this author to formulate models that were not based solely on statistical considerations but rather on hypotheses generated by previous epidemiological studies. For the two trial clusters in which no non-swimmer developed gastroenteritis, 0.5 was added to each cell in order to avoid numerical problems caused by division by 0. The logistic regression analysis was carried out only on the data pertaining to highly credible symptoms since current US EPA bacteriological water quality criteria were derived directly from these data.

Table 9.4 – Logistic regression of the log odds of gastroenteritis among swimmers (see text for further details of the analysis).

Model	Likelihood Ratio x^2	P	Log_{10} Enterococci Density + 1[a]	Location[b]	Interaction[c]	Constant
1.	28.27	< 0.0001	0.4260 (5.45)			-4.2576
2.	23.92	< 0.0001	0.3145 (3.07)	(1) -0.4684 (-3.51) (2) 0.1407 (0.75)		-3.8835
3.	18.44	= 0.0001	-0.0218 (-0.17)	(1) -1.7636 (-5.27) (2) -1.6177 (-1.66)	(1) 0.8475 (4.17) (2) 1.5915 (1.52)	-3.2388

[a] Entered into the model as a continuous variable.
[b] (1) = New York City, (2) = Boston. Reference group is Lake Pontchartrain.
[c] (1) = New York City x Log_{10} (Enterococci Density + 1),
(2) = Boston x Log_{10} (Enterococci Density + 1).

The results of the logistic regression analysis are shown in Table 9.4. The first variable entered was enterococci density. Since enterococci density was entered into the model as a continuous variable, it was necessary to apply a Log_{10} transformation to the observed enterococci density in order to insure linearity in its logit. In addition, 1 was added to each enterococci density prior to transformation so that evaluation of the model could be conducted at zero enterococci density.

Model 1 in Table 9.4 shows enterococci density to significantly improve the model ($P < 0.0001$). The next variable entered into the model was constructed to test for possible effects of marine vs brackish waters, as well as possible differences among marine water locations. Since the EPA study was conducted at two marine water locations (New York City and Boston) and one brackish water location (Lake Pontchartrain), it was possible to construct a set of indicator variables that could access the effect of each of the locations on Log odds of gastroenteritis among swimmers. These indicator variables were constructed using brackish water (Lake Pontchartrain) as the reference category. Model 2, Table 9.4 shows that the addition of these indicator variables to model 1 significantly improved the model (P < 0.0001). Inspection of the regression coefficient for Log_{10} (enterococci density + 1) in models 1 and 2 shows that location does not confound the relationship between enterococci density and gastroenteritis among swimmers to any appreciable degree. Possible interaction between the effects of enterococci density and location was then assessed. Significance of the interaction terms would demonstrate different relationships between enterococci density and gastroenteritis among swimmers at each of the three study locations. Model 3 Table 9.4 shows that introduction of this interaction term significantly improved the model (P = 0.0001). The overall goodness of fit of model 3 was assessed using both the Pearson x^2 test and the Hosmer-Lemeshow x^2 test. The Pearson x^2 test yielded a $x^2 = 18.31$ with 12 degrees of freedom (P = 0.11). The Hosmer-Lemeshow $x^2 = 0.477$ with 2 degrees of freedom (P = 0.79). These results indicate that model 3 fits the logistic model quite adequately.

Model 3, Table 9.4 was then used to estimate the probability that a swimmer would acquire gastroenteritis at varying enterococci densities at each of the locations studied. Manipulation of the regression coefficients yielded the following mathematical relationships among the locations studied:

Lake Pontchartrain:

L_n odds of gastroenteritis among swimmers =

- 0.0218 Log (enterococci density + 1) - 3.2388

New York City:

L_n odds of gastroenteritis among swimmers =

0.8257 Log (enterococci density + 1) - 5.0024

Boston:

L_n odds of gastroenteritis among swimmers =

1.5697 Log (enterococci density + 1) - 4.8565

The significance of the regression coefficient for the above 3 equations are $P = 0.86$ for Lake Pontchartrain, $P =< 0.0001$ for New York City and $P = 0.13$ for the Boston location.

Figure 9.3 is a plot of the probability of a swimmer acquiring gastroenteritis vs increasing enterococci density at each study location based on the 3 mathematical relationships shown above. Inspection of Figure 9.3 shows the probability of acquiring swimming-associated gastroenteritis is much greater at the Boston location relative to the other two study locations at any given enterococci density. Figure 9.3 further shows that while the probability of a swimmer acquiring gastroenteritis rises with increasing enterococci density at the marine water locations, this probability remains essentially constant at the Lake Pontchartrain study location.

Therefore, while enterococci densities are predictive of swimming-associated gastroenteritis at the marine water locations, they are not predictive of gastroenteritis at the brackish water location. Further, Figure 9.3 shows significant differences in the mathematical relationship between swimming-associated gastroenteritis and enterococci density at the two marine water locations used in the EPA study. Therefore, not only was the magnitude of the association related to the specific study location sampled, but the possibility exists that the very existence of any relationship between enterococci density and gastroenteritis may also be site specific. It should be noted that the highest enterococci densities observed at the New York City and Boston locations were 12/100 ml and 298/100 ml respectively. Caution must therefore be exercised when extrapolating beyond these densities in the curves depicted in Figure 9.3 (the highest enterococci density observed at Lake Pontchartrain was 711/100 ml). This caution does not detract from the fact that combining the data from these three study locations prior to analysis, as was done in the EPA study, was clearly inappropriate.

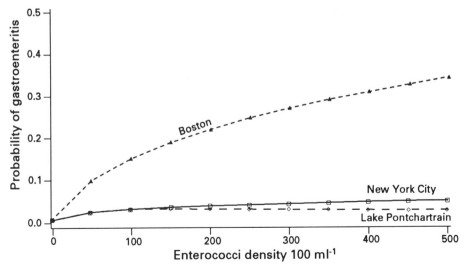

Figure 9.3 – Probability of gastroenteritis among swimmers versus enterococci density.

DISCUSSION

Current US EPA bacterial water quality criteria governing marine recreational water state:

> *Based on a statistically sufficient number of samples (generally not less than 5 samples equality spaced over a 30-day period), the geometric mean of the enterococci densities should not exceed 35 per 100 ml. No sample should exceed a one-sided confidence limit using the following as guidance: designated bathing beach - 75% C.L.; moderate use for bathing 82% C.L.; light use for bathing 90% C.L.; and infrequent use for bathing - 95% C.L. (all confidence limits are to be based on a site specific log standard deviation, or if site data are insufficient to establish a log standard deviation, then using 0.7 as the log standard deviation)*

(USEPA, 1986)

The maximum allowable geometric mean enterococci density of 35 per 100 ml was obtained by assuming an *acceptable* swimming-associated rate of gastroenteritis of 19 per 1000 swimmers, inserting this rate as the rate difference in the linear regression equation corresponding to *Highly Credible* GI symptoms derived by the EPA study, and back solving for the corresponding enterococci density predicated by the equation. This *acceptable* rate of swimming-associated gastroenteritis is equal to the estimated rate of illness at 200 faecal coliform organisms per 100 ml which was the criteria governing marine recreational waters prior to enactment of current federal criteria based on enterococci density (USEPA, 1976). Since this report has shown that the regression equation used to arrive at the current maximum allowable mean enterococci density incorporated within it three very different mathematical relationships relating gastroenteritis to enterococci density, the validity of the use of this maximum allowable geometric mean density becomes quite questionable. Using the results of the re-analysis presented in this report (Model 3, Table 9.4), the probability of gastroenteritis to the individual swimmer at 35 enterococci per 100 ml would be 0.024, 0.082, and 0.036 at New York City, Boston, and Lake Pontchartrain, respectively. The large difference in risk between the two marine water locations (New York City and Boston) casts serious doubt on validity of the use of a single maximum allowable geometric mean enterococcus density for all marine water locations as put forth by current federal criteria.

It is interesting to look more closely at how the actual probability of gastroenteritis to the individual swimmer rises as a function of current Federal criteria. Table 9.5 shows the probability of gastroenteritis to the individual swimmer derived from the results of the logistic regression analysis (Model 3, Table 9.4) at two and three times the current maximum allowable mean enterococci density at each of the study locations. Inspection of Table 9.5 shows that, at the New York City location,

Table 9.5 – Probability of gastroenteritis to an individual swimmer at the New York City, Boston, and Lake Ponchartrain study locations at 1, 2, and 3 times maximum geometric mean enterococci densities as put forth by current US Environmental Protection Agency guidelines for marine recreational waters (see text for further details).

Enterococci Density/100 ml	LOCATION		
	New York City[a]	Boston[a]	Lake Pontchartrain[b]
35 (maximum allowable)	0.024	0.082	0.036
70 (2x maximum allowable)	0.030	0.124	0.036
105 (3x maximum allowable)	0.034	0.157	0.036

[a]Denotes marine water location.   [b]Denotes brackish water location.

maximum allowable mean enterococci densities can be increased significantly with little increase in risk. Table 9.5 further shows that, although risk increases somewhat faster at the Boston location, maximum allowable mean enterococci densities could also be increased at this location without large increases in risk to the individual swimmer. These observations have obvious implications pertaining to unnecessary beach closures.

Current USEPA guidelines governing marine recreational water require the use of confidence intervals to arrive at maximum allowable single sample enterococci density. Such use of confidence intervals requires that the standard deviation of enterococci density be determined at each specific location or use of the standard deviation derived from the EPA study. Deriving site specific estimates of the standard deviation will prove cumbersome and costly while there is little reason to suppose that the EPA estimate will adequately estimate other site specific standard deviations. Standards based on estimates of risk to the individual swimmer would not require such use of confidence intervals and, therefore, would prove less cumbersome and less costly. All that would be required is to arrive at some *acceptable* level of risk to the individual swimmer. Marine recreational water quality standards based on estimates of risk to the individual swimmer derived from statistical methodology similar to that used in the re-analysis presented in this report would therefore simplify the process of defining and implementing future water quality criteria.

Based on the serious methodological and analytical weaknesses incorporated in the EPA study as shown by this report, it would be premature to conclude that health effects have been quantified sufficiently to support the continued use of current federal bacteriological criteria governing marine recreational waters. The practical significance of this finding cannot be over stressed. Currently, most local health departments still use recreational water quality standards based on the use of coliform organisms. To require a change of indicator organism at this point would be inappropriate and would result in large but unnecessary costs. This is especially true in light of the fact that the current *acceptable* level of risk to the swimmer remains the same under previous Federal criteria that used faecal coliforms as the indicator organism of choice. Perhaps of more importance is the fact that the re-analysis presented in this report questions the appropriateness of the use of a single maximum allowable mean enterococci density to govern all marine recreational water locations in the United States.

REFERENCES

BMDP (1988) *University of California: BMDP Statistical Software.* Los Angeles, CA: University of California Press.

Breslow, N.E. and Day, N.E. (1980) *Statistical methods in cancer research. Volume 1: The analysis of case-control studies.* New York: Oxford University Press.

Cabelli, V.J., Dufour, A.P., McCabe, L.J. (1982) Swimming-associated gastroenteritis and water quality. *American Journal of Epidemiology 115*, 606-616.

Ferley, J.P., Zmirou, D. and Balducci, B.E. (1989) Epidemiological significance of microbiological pollution criteria for river recreational waters. *International Journal of Epidemiology 18*, 198-205.

Hoffman, R.E. (1984) The use of epidemiological data in the courts. *American Journal of Epidemiology 120*, 190-202.

Kleinbaum, D.G., Kupper, L.L. and Muller, K.E. (1988) *Applied regression analysis and other multivariable methods.* Boston, MA: PWS-KENT Publishing Company, 1988.

PHLS (1953) Sewage contamination of coastal bathing waters in England and Wales. *Journal of Hygiene, Cambs. 57*, 435-472.

Rothman, KJ. (1986) *Modern Epidemiology.* Boston, MA: Little, Brown and Company.

Seyfried, P.L., Tobin, R.S., Brown, N.E., Ness, P.F. (1985) A prospective study of swimming-related illness II - morbidity and the microbiological quality of water. *American Journal of Public Health 75*, 1071-1075.

Stevenson, A.H. (1953) Studies of bathing water quality and health. *American Journal of Public Health 43*, 529-538.

U.S. Army Corp of Engineers (1989) Personal Communication: New Orleans, LA.

U.S. Environmental Protection Agency (1972) Quality criteria for water. Washington, DC: U.S. Government Printing Office.

USEPA (1983) U.S. Environmental Protection Agency: *Health effects criteria for marine recreational waters. USEPA Publ. No. EPA3 600/1-80-031.* Triangle Park, NC: Health Effects Research Laboratory.

USEPA (1986) U.S. Environmental Protection Agency: *Ambient water quality criteria for bacteria - 1986. USEPA Pub. No. [EPA] 4401584-002.* Washington, DC: Office of Regulations and Standards.

Velz, C.J. (1970) *Applied Stream Sanitation.* New York: John Wiley and Sons, .

Chapter 10

RECENT EPIDEMIOLOGICAL RESEARCH LEADING TO STANDARDS

David Kay and Mark Wyer
Director and Research Fellow; Centre for Research into Environment and Health, University of Wales, Lampeter SA48 7ED

INTRODUCTION

The quality of marine recreational waters is a matter of intense public and scientific concern in many nations (Oldridge, 1992 this volume; Jones, 1981; HMSO, 1984; 1985a,b,c; Galbraith *et al.*, 1987; Garrett, 1987; Jones and Kay, 1989; DoE, 1989a,b; 1990, GESAMP, 1990). In the United Kingdom significant efforts have been made to present data to the public on the quality of recreational waters (Figure 10.1). This concern and agency response has not been equalled by parallel issues such as riverine or drinking water quality. Public concern stems from the possible health effects of bathing in sea water contaminated by sewage. This problem has received considerable research attention during the last decade with little concrete progress to facilitate the definition of appropriate standards for the management and control of perceived risks. This paper examines the current state of knowledge on the possible health effects of bathing in sewage contaminated coastal waters and presents the results of a novel research protocol designed to overcome many of the pitfalls identified by Jay Fleisher in the previous chapter.

EARLY INVESTIGATIONS AND WATER QUALITY STANDARDS

The first attempts to quantify this problem in the modern period began in the 1950s with studies in the USA and the UK (Stevenson, 1953; PHLS, 1959; MRC, 1959). Stevenson (1953) examined bathing sites on Lake Michigan, the Ohio River, Tacoma Park Pool (a chlorinated recirculating fresh water pool) and the tidal waters of Long Island Sound. The approach adopted was to select comparative bathing

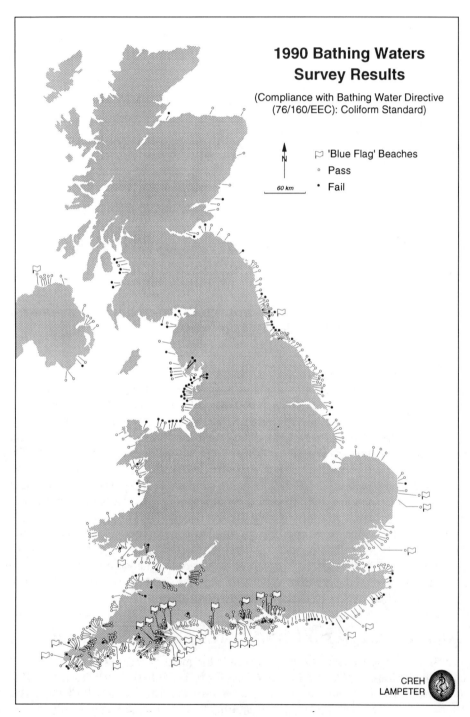

Figure 10.1 – Bathing water quality in England and Wales 1990.

waters at each site which had *poor* and *good* water quality. Water quality was measured using total coliform organisms only and the health effects of bathing were quantified using a diary sheet to acquire data on bathing history and symptomatology from the adjacent resident populations. The marine site used beaches at New Rochelle (median total coliforms 610 100ml^{-1}) and Mamaroneck (median total coliforms 253 100ml^{-1}). The participants in this investigation averaged 60 days of health and activity observations and 4,590 individuals participated in the New Rochelle study with 4,930 at Mamaroneck. A total of 3,300 illnesses were reported. At the more polluted site these numbered 5.3 per 1,000 person days whilst at the less polluted site, of Mamaroneck, a higher rate of 6.2 per 1,000 person days was reported. These marine morbidity rates were lower than the comparable data for all the freshwater locations examined (i.e. Great Lakes 7.1-8.3 per 1,000 person days and Ohio River-Pool 8.8-13.8 per 1,000 person days). It is interesting to note that the highest incidence of illness, 13.8 per 1,000 person days was found in the pool which demonstrated the lowest total coliform density (i.e. zero total coliforms 100ml^{-1}). Significant excess perceived disease incidence rates were observed on the Ohio River and Lake Michigan with bathing water median total coliform levels of 2,700 100ml^{-1} and 2,300 100ml^{-1} respectively. Stevenson (1953) concluded that:

> *sufficient evidence is available to indicate that some of the strictest bacterial quality requirements now existent might be relaxed.*

The PHLS (1959) studies in the United Kingdom adopted a different approach to the problem. A *retrospective* research design was used in which the PHLS committee, led by Dr Brendan Moore, examined notified cases of poliomyelitis and enteric fever over a five year period between 1953 and 1958. The reasons for restricting the study to these two diseases were not explained beyond the statement on page 458 that:

> *The committee was particularly anxious to assess the risk of contracting enteric fever or poliomyelitis through bathing in sewage polluted sea water.*

Given the low incidence of these notifiable illnesses in the population, the decision to focus on this narrow set of possible infections effectively restricted the research design to the retrospective protocol adopted and precluded the more *prospective* methodology of Stevenson (1953). The aim of the PHLS committee was therefore to examine the bathing history of all notified cases of the two ailments and a suitable matched control group. Only 4 cases of paratyphoid fever, notified during the three years from 1956-1958, could be related to bathing in sewage polluted water or playing on polluted beaches. More *anecdotal* evidence of paratyphoid infection associated with sewage polluted sea waters was presented on two cases in 1950 and 1952 and a further 4 cases in the period 1946-1949. An analysis of notified poliomyelitis cases caused the committee to suggest that:

for patients suffering from poliomyelitis a history of having bathed during the 3 weeks preceding the onset of symptoms is probably irrelevant as a causal factor.

The PHLS committee concluded (page 467):

it could be argued that bathing waters with median coliform counts of greater than 10,000 per 100ml occasionally cause paratyphoid fever, and that a standard of this order can be justified on health grounds

and (page 468-9)

(i) *That bathing in sewage-polluted sea water carries only a negligible risk to health, even on beaches that are so fouled as to be aesthetically very unsatisfactory.*

(ii) *That the minimal risk attending such bathing is probably associated with chance contact with intact aggregates of faecal material that happen to have come from infected persons.*

(iii) *That the isolation of pathogenic organisms from sewage-contaminated sea water is more important as evidence of an existing hazard in the populations from which the sewage is derived than as evidence of a further risk of infection in bathers.*

(iv) *That since a serious risk of contracting disease through bathing in sewage-polluted sea water is probably not incurred unless the water is so fouled as to be aesthetically revolting, public health requirements would seem to be reasonably met by a general policy of improving grossly insanitary bathing waters and of preventing so far as possible the pollution of bathing beaches with undisintegrated faecal matter during the bathing season.*

EARLY STANDARDS

Both the PHLS (1959) and Stevenson (1953) investigations have profoundly affected policy in their respective countries. The US standards proposed in 1968 by the National Technical Advisory Committee (NTAC) of the Department of the Interior were based directly on Stevenson's studies (NTAC, 1968). The total coliform level of 2,300 100ml^{-1}, associated with significant excess disease incidence was, converted to a faecal coliform level using the ratio between the two species experienced in the Ohio River where the studies took place. At this location 18% of the coliforms were *faecal* and this suggested a median standard of 414 faecal coliforms 100ml^{-1} defined the point at which a detectable risk was evident. To provide an adequate safety margin the median concentration of 200 faecal coliforms 100ml^{-1} was therefore proposed and, in addition, the NTAC felt that recreational waters should not present a health risk in more than 10% of the time. Hence, the standard proposed was as follows (USEPA, 1986:2):

> *Fecal coliforms should be used as the indicator organism for evaluating the microbiological suitability of recreation waters. As determined by the multiple-tube fermentation or membrane filter procedures and based on a minimum of five samples taken over not more than a 30-day period, the fecal coliform content of primary contact recreation waters shall not exceed a log mean of 200/100ml, nor shall more than 10 percent of total samples during any 30 -day period exceed 400/100ml.*

The weak epidemiological base for this standard was noted by many workers (Henderson, 1968; Moore, 1975, Cabelli *et al.*, 1975) who pointed to the selective use of the Lake Michigan data, the inclusion of all illnesses, some of which might not be related to sewage pollution, the poor definition of *bathing* and the lack of beach-going but non-swimming control groups. Indeed the National Academy of Sciences (1972:30) declined to recommend a water quality criterion for recreational waters due to the lack of adequate epidemiological information.

United Kingdom policy was solidly based on the conclusions of the PHLS investigations for a considerable period into the mid-1980s. This confidence that the PHLS research had effectively proven the absence of epidemiological risk at the overwhelming number of UK bathing waters could not be justified on the available evidence. Indeed, it would be scientifically invalid to infer from negative findings in a retrospective epidemiological study that a relationship does not exist (Dufour, 1982). Notwithstanding this dubious scientific foundation a water industry spokesperson, addressing a Parliamentary enquiry into sewage pollution of coastal waters in 1984 stated that:

> *A committee of the MRC conducted epidemiological studies relating to polio and enteric fever between 1955 and 1959 and it was their conclusion that there was no significant risk to health 'unless waters were so fouled as to be aesthetically revolting' This conclusion was accepted by the United Kingdom government and has been the basis of national policy since its publication.*
>
> (HMSO 1985c:25)

The PHLS findings were further used in the United Kingdom to justify a less than enthusiastic implementation of the European Community (EC) Bathing Waters Directive (EEC/76/160) defined in Table 10.1 (Moore, 1975, 1977; Barrow, 1981; Kay and McDonald, 1986a,b). This scepticism was understandable in view of the fact that the EC standards could not claim, even the inadequate, epidemiological base provided for the NTAC/USEPA standards in the United States.

Table 10.1 – Recreational water quality standards.

NORTH AMERICAN STANDARDS

AGENCY	REGIME	Faecal Coliform Standard
TORONTO HEALTH DEPARTMENT	DAILY	GM<100.100ml^{-1} NO SAMPLE TO EXCEED 400 .100ml^{-1}
CANADIAN FEDERAL	5/30 DAYS	GM<200 100ml^{-1} RESAMPLE IF ANY SAMPLE EXCEEDS 400 100ml^{-1}
U.S.E.P.A.	5/30 DAYS	GM<200 100ml^{-1} <10% ONLY TO EXCEED 400 100ml^{-1}

GM=geometric mean

EUROPEAN STANDARDS

	total coliform .100ml^{-1} (fortnightly sampling)	E. coli .100ml^{-1}
Guide (recommended) 80% of samples should not exceed this figure.	500	100
Imperative (mandatory) 95% of samples should not exceed this figure.	10,000	2,000

The only standards claiming an epidemiological basis (from the work of Stevenson, 1953) were founded on data derived wholly from fresh water recreation sites. All studies on marine waters conducted by Stevenson and the PHLS had failed to find an association between water quality and the morbidity amongst bathers. This scientific vacuum was unacceptable in view of the significant expenditures required to effect water quality improvements in the coastal zone. It was clear by the early 1970s, therefore, that an urgent requirement for high quality epidemiological information existed.

PROSPECTIVE EPIDEMIOLOGICAL STUDIES AND RECENT WATER QUALITY STANDARDS

In 1972 the USEPA commenced a series of integrated studies designed to define the epidemiological risks associated with recreation at marine and freshwater beaches and offer guidance on appropriate indicator systems which would define these risks and offer a management control strategy through the design of appropriate standards (Cabelli *et al.*, 1975, 1982; USEPA, 1986).

The study design required that; (i) discrete trials were conducted on weekends only to reduce the confounding factors of multiple exposures. (ii) Demographic information was gathered by an initial beach interview. (iii) Bathing activity was ascertained at the initial beach interview and bathers strictly defined as those with wet hair, a non-swimming control group was also recruited at the beach and where possible family groups were recruited. (iv) Reminder letters were sent to all recruits one or two days after the initial beach interview. (v) Eight to 10 days after the beach contact recruits were contacted by telephone to determine their symptomatology and demographic characteristics which were ascertained during a telephone questionnaire interview. (iv) In the absence of medical and clinical follow-up examinations, there was an obvious requirement to define those perceived symptoms which were considered medically credible. This was done by defining Highly Credible Gastrointestinal symptoms (HCGI) as vomiting, diarrhoea accompanied by fever or that were disabling; and nausea or stomach-ache accompanied by fever. These studies were conducted at fresh water and marine recreation sites and reported by Cabelli *et al.* (1983, 1982) and Dufour (1984).

The USEPA prospective studies demonstrated an excess rate of gastrointestinal symptoms in swimmers using the *barely acceptable* beaches when compared to non-swimmers and bathers using *relatively unpolluted* beaches. Cabelli *et al.* (1982) presented log-linear bivariate relationships between indicator concentrations in the bathing water and the excess symptom attack rates experienced by the bather populations. From these dose-response relationships Cabelli *et al.* (1983) suggested that appropriate indicator concentrations in marine waters could be defined on the basis of acceptable symptom attack rates in the recreator population. The enterococci health effects relationship for marine waters is shown in Figure 10.2.

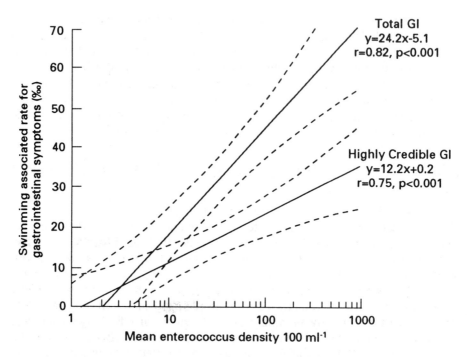

Figure 10.2 – Dose-response relationships produced by the work of Cabelli *et al.* (1982).

RECENT STANDARDS

Cabelli *et al.* (1983) suggest a number of problems which their investigations had not identified. First, that little data were available on the nature of the main aetiological agent(s) in the transmission of gastrointestinal symptoms contracted from coastal bathing. Second, the requirement to identify human specific and environmentally resistant microbial indicator species and, finally, the necessity for a separate criterion for fresh recreational waters. Cabelli's work represented the first credible evidence of an excess disease risk from bathing in marine recreational waters. The USEPA (1986:8-10) standards based on these investigations recommend the following compliance protocol.

(i) Samples should be collected during dry weather to establish *steady state* conditions avoiding storm water effects.

(ii) Sampling should be at weekly intervals during high use periods and bi-weekly or even monthly during low use periods.

(iii) Compliance is based on a defined geometric mean value and a single value which no sample should exceed.

(iv) Calculation of the geometric mean value is based on at least five samples collected at equal temporal spacings over a 30 day period.

(vi) Calculation of the single value is based on a knowledge of the unique site $\log_{10}$ standard deviation and the usage rate of the location. Standard deviation values of 0.4 for fresh water *E. coli* and enterococci and 0.7 for marine water enterococci can be assumed where no site specific calculation is possible for this parameter due to insufficient data.

It is not intended that this standard should be implemented immediately but that States should evaluate the impact of these recommendations on bathing water compliance against their existing sampling regimes. However, it is estimated that application of the current geometric mean value of 200 faecal coliforms 100ml^{-1} would result in 8 illnesses per 1000 bathers at fresh water sites and 19 per 1000 bathers at marine bathing locations. This is termed the *acceptable swimming associated gastroenteritis rate*. The calculated geometric mean values for fresh and marine waters is presented in Table 10.2 together with an explanation of the different use categories specified. These standards present considerable scope to accommodate both the intensity of recreational use and the variable standard deviation experienced at different recreation sites. The latter parameter of the bacterial probability density function provides an important predictor of the numbers of single sample non-compliance results and hence the probability of unacceptable risk. This method of compliance assessment is similar to the use of laboratory control charts, now a familiar element in standard AQC procedures.

Judgement on the success of these new USEPA standards must await further information on their practicality and ease of use by the competent authorities at State level. However, they offer the first attempt to formulate flexible and epidemiologically based criteria designed to limit the health risk to *acceptable* levels. The flexibility in this system should overcome many of the problems caused in the present USEPA criteria which can produce many non-compliance results because of the criterion that not more than 10% of samples should exceed 400 faecal coliforms 100ml^{-1}. This value is much lower than the 90%ile point of a $\log_{10}$-normal distribution with a standard deviation of 0.7 and it therefore triggers non-compliance even where the geometric mean value is well below 200 faecal coliform organisms 100ml^{-1} (Brown et al., 1987a,b; Ellis and Stanfield, 1987, ; Kay et al., 1980).

RECENT EPIDEMIOLOGICAL INVESTIGATIONS

A significant problem for the competent authorities, now faced with implementation of the USEPA (1986) standards, is that these are based on an epidemiological

Proposed USEPA Standards

Freshwater standards (values 100 ml^{-1})
assuming acceptable swimming associated GI = 8 ‰

Indicator	Geometric mean	Single Sample Allowable Density			
		Designated Beach area	Moderate Contact	Light Contact	Infrequent Contact
Enterococci	33	61	89	108	151
Faecal coliform	126	235	298	406	576

Marinewater standards (values 100 ml^{-1})
assuming acceptable swimming associated GI = 19 ‰

Indicator	Geometric mean	Single Sample Allowable Density			
		Designated Beach area	Moderate Contact	Light Contact	Infrequent Contact
Enterococci	35	104	158	276	500

Notes:
The single sample allowable density is based on no sample exceeding the following upper confidence limits:

> Designated Beach   75%
> Moderate Use   82%
> Light Use   90%
> Infrequent Use   5%

based on knowledge of site specific $\log_{10}$ standard deviation (or values of 0.4 for freshwaters and 0.7 for marinewaters where the $\log_{10}$ standard deviation is unknown) and a sample size of not less than 5 samples equally spaced over 30 days.

Also geometric mean densities should not exceed the following criteria:

> Freshwater   enterococci   33 100ml^{-1}
>   faecal coliform   126 100ml^{-1}
>
> Marinewater   enterococci   35 100ml^{-1}

Table 10.2 – Criteria for indicator for bacteriological densities (USEPA 1986).

protocol which has increasingly been questioned. For example, Fleisher (Chapter 9 in this volume) has re-analysed the data on which the 1986 USEPA standards are based. He has demonstrated that the mathematical relationship between bather morbidity and enterococcus concentration assumed by the USEPA is of dubious quality. In consequence, the 1986 USEPA standards do not have a firm scientific foundation.

Furthermore, he has identified significant bias in the USEPA studies due to measurement error in the bacterial enumerations which was not accounted for in the original analyses (Fleisher, 1990b).

In a methodological review paper, Jones et al. (1990) concluded that;

> *Little firm data exist on the epidemiological significance of existing disposal practices and the established prospective methods, developed by North American workers in this area, have not provided a scientifically robust epidemiological research protocol. This leaves a scientific vacuum for those with operational and policy responsibilities in this important and sensitive area of water resource management.*

This harsh judgement on the USEPA protocol was based on an examination of replications of the methodology and recent investigations by Lightfoot (1989) and the New Jersey Health Department (1988, 1989). Results of these studies are summarised in Table 10.3. Jones et al. (1990) considered that the USEPA had, in effect, designed a protocol which had proven to be unreplicable and it therefore failed the first test of an acceptable scientific experiment. The likely reason for this failure is that the USEPA protocol is founded on *perception* data alone. The acquisition of perception data in health studies is difficult and contentious. It is for this reason that the most rigourous epidemiological methods must be employed to minimise effects caused by interviewers and respondents. In the most recent implementation of the USEPA protocol, Lightfoot (1989) specifically tested for such effects and she found that;

> *there was no evidence to suggest that bacterial count contributed to the prediction of illness in swimmers. Instead, age, contact person, and interviewer, most frequently tended to be important.*
> (Page iv)

> *There is little evidence from the present study to support the belief that the bacterial water quality indicators investigated herein index the short-term risk of becoming ill from swimming*
> (Page 208)

Lightfoot makes a number of significant recommendations and comments.

Table 10.3 – Summary results of the Cabelli style and other epidemiological perception studies.

AUTHOR	DATE	NATION	FRESH/SEA	INDICATOR	R^2	SYMPTOMS
Stevenson	1953	USA	both	total coliform	NR	ENT/GI/R
Cabelli	1982	USA	both	enterococci	.56	GI
Seyfried	1985	Canada	fresh	total staphylococci	.19	R/GI
				faecal coliform	.08	
				faecal streptococci	.03	
Lightfoot	1989	Canada	fresh	age	NR	R/GI
				contact person		
				interviewer		
Cheung	1988/90	Hong Kong	sea	E. coli	.53	S/GI
				staphylococci		
El Sharkawi	1983	Egypt	sea	enterococci	.79	GI
				E. coli	.77	
Fattal	1986	Israel	sea	enterococci	NR	GI
				E. coli	NR	GI
Mujeriego*	1982	Spain	sea	faecal streptococci	NR	S/E/ENT/GI
Foulon	1983	France	sea	faecal streptococci	NR	E/S/GI
				total coliforms		
				faecal coliforms		
Ferley**	1989	France	fresh	total coliforms	.21	All + S
				faecal coliforms	.45	S
				faecal streptococci	.38	GI
				Aeromonas spp.	.26	S
				P. aeruginosa	.53	S

List of Symptoms

NR = not reported
E = eye symptoms
S = skin complaints
GI = gastrointestinal symptoms
ENT = ear nose and throat symptoms
R = respiratory illness

R^2 = Coefficient of determination (a measure of the degree of explanation in the dependent variable provided by the independent variable)

Sources All named authors and Shuval

* a cross sectional study
** a retrospective study, the highest R^2 values for each parameter is reported here

> *It is possible that the utilisation of medical and laboratory confirmation might have altered the results which were based on the reporting of illness.*
> (Page 206) and

> *future investigators will be well advised to attempt recording of duration of exposure for individuals, and to carry out more frequent water sampling each day than was possible in the present study.*
> (Page 207) and

> *Yet another potential source of bias in this study is that illness was reported by contact persons, and not confirmed by clinicians and laboratory testing. .... It may prove beneficial for investigators of swimming related illness to compare results from the two methods of reporting (i.e. contact person versus the use of clinicians and laboratory results)*
> (Pages 223-226)

These recommendations are similar to WHO (1972:13) suggestions in this area, namely;

> *Ideally, the best hope of progress in this field would seem to lie in carefully planned prospective studies in volunteer populations of adequate size. If such populations could be randomly divided into comparable groups of persons who did and did not bathe, but shared all other activities and exposures to environmental hazards, so much the better. The populations would need to be observed at close quarters by teams that included clinicians, public health workers, epidemiologists and microbiologists.*

Work along these lines has now been pilot tested during the summers of 1989 at Langland Bay, near Swansea UK (WRc, 1989,1990), and in subsequent bathing seasons at Moreton, on the Wirral, UK (4th August 1990) and at Southsea near Portsmouth UK (6th July 1991). The results of the Langland Bay investigation are outlined below. Data from the latter two studies are still under examination.

CONTROLLED COHORT EPIDEMIOLOGICAL STUDIES

The controlled cohort approach requires volunteer groups of bathers and non-bathers which share similar environmental exposure patterns except that one group bathes in seawater. The protocol used received the prior approval of the Royal College of Physicians Committee on Ethical Issues in Medicine. The pilot scale implementation of this protocol at Langland Bay involved adult volunteers, over eighteen years of age, recruited mainly from the City of Swansea. The 276 volunteer cohort was randomly split into bather and non-bather groups on arrival at the beach. Non-bathers were given a packed lunch and invited to have a picnic on the beach. The bather cohort took part in a closely supervised dip in the sea which lasted a minimum of 10 minutes. During this time the bathers were asked to immerse their heads on at least three occasions. Bathing time, location, activities and duration were recorded by the supervisors. Bathing location was defined by six marker posts at 20m intervals over a 100m stretch of beach. These markers were also used for accurate positioning of the water sampling points. Some 180 seawater samples were collected over a 3 hour period; that is, three depths at each marker (surf, 30cm and chest depth) sampled at 20 minute intervals plus quality control samples and boat samples at 50m off-shore. Following the bathing and immersion period all the volunteers were given a packed lunch which they ate on the beach before dispersing.

The objective of this experiment was to simulate, as far as possible, normal beachgoing activity of bathers and non-bathers. Both cohort groups experienced similar epidemiological risks in travelling to the beach and consuming a packed lunch, the only differential risk factor between the two being the direct exposure to seawater.

The bacteriological quality of the seawater during the test period 12.30pm to 3.00pm on 2nd September 1989 was relatively good in UK and EC terms (Table 10.4). Fifteen samples were also analysed for rotaviruses and enteroviruses. Three samples analysed for viruses were positive for rotaviruses (4, 4 and 8 ff.10 l^{-1}) and one of the fifteen for enteroviruses (2 pfu. 10 l^{-1}).

Table 10.4 – Summary statistics, all data, Langland Bay 2/9/89.

Variable	Mean	Standard Deviation	Min	Max	N*
Log_{10} Total coliform bacteria	1.567	0.759	-1.00	3.16	180
Log_{10} Faecal streptococci	1.501	0.530	-1.00	2.29	180
Log_{10} Pseudomonas aeruginosa	-0.758	0.629	-1.00	2.30	180
Log_{10} Faecal coliform bacteria	1.295	0.850	-1.00	3.12	180

Note: all Log_{10} values are log_{10} (count (per 100 ml) + 0.1)
* = Number of samples from which the mean and standard deviations were calculated

Three days before, during and three days after the test day (i.e. 2nd September 1989) the volunteers underwent a series of three questionnaire interviews. At each interview a questionnaire was completed by a trained interviewer. Most of the interviewers were environmental health officers from the City of Swansea. A fourth questionnaire was distributed three weeks after the trial and completed by the volunteers themselves who were, by that stage, familiar with the consistent questionnaire design. The questionnaires were designed to acquire the best quality data on the volunteers' perception of illness at each stage in the experiment. Questions included inquiry on dietary habits, social background, symptoms and severity of recent illness, chronic illness, type and frequency of other recreational water exposure and travel history. The questionnaire set was designed by medical epidemiologists and resulted in over 600 items of data on each volunteer.

At the pre-trial and post-trial interviews the volunteers were examined by medical staff who took ear and throat swabs. Volunteers presented a faecal sample at each of these interviews and a third faecal sample was presented with the final (three-week) questionnaire form. Ear and throat swabs were analysed for five bacterial species. Faecal samples were analysed for cysts, ova, parasites and enteroviruses as well as the bacterial pathogens *Campylobacter* spp. and *Salmonella* spp.

PERCEIVED SYMPTOM ATTACK RATES

Crude perceived symptom attack rates for bathers and non bathers are shown in Table 10.5. The significance of the bather/non-bather differentials is presented in Table 10.6. Crude perceived symptom attack rates range from 238/1000 to zero and they are considerably higher than those reported in previous prospective perception studies which used telephone follow-up of the subjects. This differential is to be expected and reflects the underestimate of actual symptom attack rates when health data are acquired from respondents reporting on whole family groups by phone rather than in a one-to-one interview solely by a, necessarily brief, telephone interview in which time constraints and domestic distractions impair the recall of a respondent who might be reporting on several volunteers.

Significant bather/non-bather differences in perceived illness were observed for sore throat, ear symptoms and eye symptoms three days after the bathing event and for diarrhoea three weeks after the exposure day. Bathers reporting symptoms were compared with bathers not reporting symptoms to investigate the possible influence of water quality on the perceived symptom attack rates. Geometric mean water quality experienced by bathers reporting a symptom did not differ significantly from that of bathers not reporting a symptom at any of the three sampling depths and for any of the bacterial indicators enumerated

CLINICAL SYMPTOM ATTACK RATES

Crude symptom attack rates for the throat and ear swabs are presented in Table 10.7 and the significance of difference between bathers and non-bathers is quantified in Table 10.8. No significant difference in water quality could be identified between that experienced by bathers whose ear and throat swabs indicated infection and those with negative results. The results of the faecal sample analysis are presented in Table 10.9. These results present so few positive determinations that statistical analysis of bather/non-bather differentials would not be appropriate.

Table 10.5 – Crude perceived symptom attack rates per 1,000 for the a) bather cohort and b) non-bather cohort.

a) Symptom	On the day	3 days after	3 week after
Fever	22.7	23.3	114.5
Headache	83.3	165.4	250.0
Aching limbs	45.4	78.1	83.3
Chest pains	7.6	0.0	22.7
Dry cough	37.9	62.5	106.9
Productive cough	53.0	46.5	84.0
Sore throat	53.4	156.3	204.5
Ear inflamation	15.2	39.1	37.9
Eye inflamation	15.3	62.0	30.3
Breathing difficulty	0.0	15.6	0.0
Blurred vision	0.0	0.0	15.3
Loss of appetite	15.2	38.8	90.9
Indigestion	0.0	23.3	31.3
Diarrhoea	15.2	62.0	121.2
Nausea	22.9	38.8	56.0
Vomiting	7.5	15.5	15.2
Lassitude	7.5	23.3	114.5
Dizziness	7.5	15.6	60.6
Skin rash	30.3	54.7	22.7
Credible G I	37.9	100.8	169.2
Diarrhoea or nausea	37.9	100.8	169.2
Ear or eye or throat	76.3	212.6	238.5
Ear or throat	68.2	181.1	223.1

b) Symptom	On the day	3 days after	3 weeks after
Fever	7.5	15.0	75.2
Headache	101.5	135.3	174.2
Aching limbs	77.5	75.2	90.2
Chest pains	0.0	15.0	22.7
Dry cough	31.0	52.6	82.7
Productive cough	77.5	90.2	67.7
Sore throat	3.8	75.2	127.8
Ear inflamation	15.5	0.0	7.5
Eye inflamation	7.5	7.5	22.6
Breathing difficulty	15.5	15.0	15.0
Blurred vision	0.0	7.5	7.5
Loss of appetite	7.5	15.0	67.7
Indigestion	23.3	7.5	37.6
Diarrhoea	0.0	52.6	37.6
Nausea	0.0	7.5	67.7
Vomiting	0.0	15.0	15.0
Lassitude	7.8	30.0	53.4
Dizziness	0.0	15.0	30.0
Skin rash	7.8	45.1	15.0
Credible G I	0.0	75.2	101.6
Diarrhoea or nausea	0.0	60.2	101.6
Ear or eye or throat	54.3	82.7	148.4
Ear or throat	46.5	75.2	132.8

Table 10.6 – Significance values (p) for Chi² analysis of significance between bather and non-bather perceived symptom attack rates.

Symptom	On the day	3 days after	3 weeks after
Fever	0.32 +	0.49 +	0.30
Headache	0.62	0.50	0.16
Aching limbs	0.28	0.93	0.80
Chest pains	0.51 +	0.26 +	0.65 +
Dry cough	0.51 +	0.73	0.54
Productive cough	0.42	0.16	0.65
Sore throat	0.57	0.04 *	0.11
Ear symptoms	0.68 +	0.03 (+)*	0.11
Eye symptoms	0.51 +	0.02 (+)*	0.51 +
Breathing difficulty	0.24 +	0.67 +	0.25 +
Blurred vision	-	0.51 +	0.50 +
Loss of appetite	0.51 +	0.23 +	0.51
Indigestion	0.12 +	0.30 +	0.49 +
Diarrhoea	0.25 +	0.74	0.01 *
Nausea	0.13 +	0.10 +	0.58
Vomiting	0.51 +	0.68 +	0.68 +
Lassitude	0.75 +	0.52 +	0.08
Dizziness	0.51 +	0.67 +	0.20 +
Skin rash	0.19 +	0.72	0.51 +
Credible gastro-intestinal (GI)	0.03(+)*	0.46	0.11
Diarrhoea or Nausea	0.03(+)*	0.23	0.11
Ear or Eye or Throat	0.47	0.00 *	0.07
Ear or throat	0.45	0.01 *	0.06

+ Fishers exact test (used when expected cell count < 5)
- untestable, 2 cells contained no positive responses
* significant at $P < 0.05$

The controlled cohort approach offers a feasible and scientifically valid means for acquiring health data. It remains, however, labour intensive and expensive in relation to the three alternative methods employed in previous studies. The utility of this method therefore depends on its potential to provide data of enhanced scientific quality when compared to the increasingly questioned epidemiological methods used in past studies.

This study demonstrates the feasibility of the controlled cohort method in generating data on the health effects of bathing which is of higher quality than previous investigations. Volunteers can be recruited and the logistical difficulties of managing a large volunteer group on a public beach can be overcome. Whilst firm conclusions on health effects should not be drawn from pilot scale investigations, initial observations from this pilot project suggest that previous studies of bathing related disease perception, using a telephone follow-up approach, might have resulted in significant under reporting of perceived symptoms in both the bather and non-bather cohort groups. It is likely, therefore, that the present debate surrounding standards based on these studies, which has been initiated by workers such as

Table 10.7 – Crude clinical symptom attack rates per 1,000 for ear and throat swabs.

(a) bather cohort only	Ear swab 1	Ear swab 2	Throat swab 1	Throat swab 2
Any determinand	336.0	206.0	377.0	543.3
Coliform bacteria	216.0	119.0	311.5	330.7
Streptococci	0.0	0.0	41.0	149.6
Streptococcus faecalis	0.0	0.0	24.6	149.6
Staphylococcus aureus	88.0	63.5	24.6	39.4
Pseudomonas aeruginosa	48.0	39.7	41.0	118.1
(b) non-bather cohort only	**Ear swab 1**	**Ear swab 2**	**Throat swab 1**	**Throat swab 2**
Any determinand	179.7	224.8	359.4	445.3
Coliform bacteria	62.5	147.3	250.0	273.4
Streptococci	7.8	0.0	54.7	109.4
Streptococcus faecalis	7.8	0.0	54.7	109.4
Staphylococcus aureus	70.3	31.0	23.4	46.9
Pseudomonas aeruginosa	31.3	77.5	31.3	70.3

Lightfoot (1989) and Fleisher (this volume), will continue until effective means of perceived symptom reporting and recording are adopted.

Clinical analysis of samples did not confirm the perceived symptoms reported by the volunteers. Whilst it is accepted practice in incidents of infection to attempt to isolate and identify the causal agent several factors could explain the absence of an association between symptoms and positive samples. These include insufficient or inappropriate microbiological determinations on clinical samples and the timing of

Table 10.8 – Significance values (p) for Chi^2 analysis of significance between bather and non-bather clinical symptom attack rates.

	Ear Swab 1	Ear Swab 2	Throat Swab 1	Throat Swab 2
Any determinand	0.0040*	0.7201	0.7721	0.1176
Coliform bacteria	0.0004*	0.5072	0.2792	0.3193
Streptococci	0.3221+	-	0.6124	0.3386
Streptococcus faecalis	0.3221+	-	0.2248+	0.3386
Staphylococcus aureus	0.6022	0.2207+	0.9525+	0.7680
Pseudomonas aeruginosa	0.4942+	0.1992	0.6796+	0.1912

+ Fishers exact test (used when expected cell count<5)
- Untestable, 2 cells contained no positive occurrences
* Significant at $P < 0.01$

Table 10.9 – Faecal sample results, number of positive occurrences.

	Salmonella spp.	*Campylobacter* spp.	Cryptosporidia
Sample 1	1	1	0
Sample 2	1+	0	0

	cyst / ova / parasite	enteroviruses	N
Sample 1	3 *Giardia lamblia*	-	266
Sample 2	3 *Giardia lamblia*+	-	260
Sample 3	1 *Giardia lamblia**	5¶	255

+ Same host(s)
* One carrier did not present a third sample, one was negative on sample 3
¶ 2 bathers, 3 non-bathers

Sample 1 - Taken prior to exposure
Sample 2 - Taken 3 days after exposure
Sample 3 - Taken 3 weeks after exposure

clinical sample collection. It is unlikely that any study of this nature will ever achieve complete definition of all microbiological agents of perceived symptoms. However, a significant increase in clinical confirmations achieved should be possible with full microbiological determinations on faecal samples collected 5-7 days after exposure and medical appraisal of symptoms. Parallel improvements in the throat and ear swab clinical investigations should be possible with the implementation of additional virological investigations. These amendments have been made in the second implementation of this research protocol which took place at Moreton (Merseyside) on 4th August 1990. Similar studies will be required before definitive statistical public health information will be available in this area of intense public concern. It is imperative that the best quality epidemiological data, on which to base policy, is available if the United Kingdom is to avoid problems with newly designed standards similar to those now evident with United States Environmental Protection Agency Criteria published in 1986.

SIGNIFICANT PROBLEMS REMAINING

The Joint Group of Experts on Aspects of Marine Pollution (GESAMP, 1990) working under a United Nations Environment Programme concluded, among other points, that microbial contamination from sewage causes many human diseases, including cholera and hepatitis A. Control requires proper design and siting of outfalls, coupled with vigorous surveillance of shellfish beds and their marketed products, as well as timely banning of contaminated seafood. Microbial contamination of sea water is also responsible for widespread outbreaks of gastro-intestinal diseases at ill-protected and crowded beaches and is a suspected cause of respiratory, ear and skin infections among bathers.

They also conclude that, at the end of the 1980s, the major causes of environmental concern in the marine environment on a global basis are coastal development and attendant destruction of habitats, eutrophication, microbial (including algal) contamination of seafood and beaches, fouling of the seas by plastic litter, progressive build-up of chlorinated hydrocarbons, especially in the tropics and sub tropics, and accumulation of tar on beaches. They also fear, due to increasing human population pressures, that the marine environment could deteriorate in the future unless strong, co-ordinated action is taken now.

A most significant question remaining in this area relates to the choice of appropriate indicator species. The bacterial indicator species in common use are not the aetiological agents of disease and they may have significantly shorter environmental survival times than the pathogens which they index.

Cabelli (1981) has suggested a viral aetiology (possibly parvovirus agents and Norwalk virus) is most probable due to the symptomatology, incubation time and self-limiting nature of bathing related gastroenteritis. However, methods for the routine isolation of this group of pathogens from environmental samples are not currently available. Indeed, over 100 different viruses are excreted in human sewage (Melnick, 1984) whilst practical methods are currently available for the isolation of only enteroviruses and rotaviruses from surface and marine waters. Evidence of swimming associated enterovirus infection with Coxsackie B5 and A16 viruses have been presented by Hawley *et al.* (1973) and Denis *et al.* (1974) respectively. However, Havelaar (1988:18) comments on the current knowledge in the field of enterovirus infections as follows;

> *The available epidemiological evidence for waterborne transmission of enteroviruses is in fact extremely limited and disputable.*

This inability to asses the viral pathogen loading of recreational waters is important in view of the longer survival of enteric viruses than bacterial indicators and pathogens and the clear epidemiological significance of viral pathogens (Wheeler, 1990; Colwell, 1987; Denness, 1987; Grimes, 1986, Tyler, 1985). In effect, direct viral enumerations of enteroviruses and rotaviruses from marine waters should be treated as an alternative index of the risk that enteric pathogens might present a disease hazard. Several workers have extended this concept of a viral index to suggest that bacteriophage enumeration could offer a practical operational water quality index of epidemiological risk (Borrego *et al.*, 1990; Havelaar and Olphen, 1989; Snowdon and Cliver, 1989; Havelaar and Pot-Hogeboom, 1988; Havelaar, 1988).

There remain significant scientific questions and consequent operational problems in the design and implementation of standards based on bacteriophage enumeration. Enumeration consistency is dependent on a uniform host cell system, environmental concentrations may be partially governed by the residence time of the sewerage system in which bacterial infection will take place. However, bacteriophages are worthy of further research attention as indicators of epidemiological risk. It is clear from the R^2 values presented in Table 10.3 that present microbial indicators fail to provide acceptable prediction of epidemiological risk. Whilst the perception data acquired and the protocols adopted might explain some element of this failure, it is likely that imperfections in the indicator systems employed will also account for a significant element in the residual variance in the regression models employed. If bacteriophage prove to be an efficient index of viral pathogen presence, then they could help reduce this component of the model stochastic variance.

It is also worth noting that the USEPA (1986) standards recognise the differential epidemiological risks associated with fresh and marine waters by the standard levels outlined in Table 10.2. Potential pathogens, such as the *Vibrio, Cryptosporidium,* and *Leptospira* have not received sufficient research attention to assess their epidemiological significance in the marine beach environment (Philipp *et al.*, 1989: West, 1989: Anon, 1985). It would be wise for future epidemiological investigations to examine these species and their epidemiological significance

CONCLUSIONS

Marine environments are a source of water recreation including bathing. At the same time they act as receiving waters and microbial reservoirs. Legislation is being established to achieve better operational control over waste discharges and the monitoring of marine bathing areas. Thus epidemiological studies are required to assess any health risk in using polluted bathing waters and in order to establish;

(a) the degree of public health risk involved in using sewage polluted water and

(b) the relationship and correlation, if any, with microbial standards and indicator bacteria or viruses.

Such studies should avoid the shortcomings identified here and by authors such as Fleisher (1990) and Lightfoot (1989). Controlled cohort investigations offer one of a number of useful ways forward. This approach has now been field tested in three replications of the method (Jones *et al.*, 1991).

REFERENCES

Anon (1985) Leptospirosis in man, British Isles 1984. *Public Health Laboratory Service, Communicable Disease Report, CDR 85/36.* London.

Anon (1988) *Fylde Coast Waters Improvements: An Environmental Statement.* North West Water, Preston, UK.

Barrow, G.I. (1981) Microbial pollution of coasts and estuaries: the public health implications. *Water Pollution Control 80(2)*, 221-230.

Borrego, J.J., Cornax, R., Morinigo, M.A., Martinez-Manzarez, E. and Romero, P. (1990) Coliphages as an indicator of faecal pollution in water. Their survival and productive infectivity in natural aquatic environments. *Water Research 24(1)*, 111-116.

Brown, J.M., Campbell, E.A. Rickards, A.D. and Wheeler, D. (1987a) *The public health implications of sewage pollution of bathing water.* Published by the University of Surrey, Robens Institute, Guildford, Surrey UK.

Brown, J.M., Campbell, E.A. Rickards, A.D. and Wheeler, D. (1987b) Sewage Pollution of Bathing Water *The Lancet 21*, 1208-1209.

Cabelli, V.J. (1981) *Epidemiology of enteric viral infections.* In Goddard, M. and Butler, M. (Eds.) *Viruses and wastewater treatment.* Pergamon, Oxford. 291-304 pp.

Cabelli, V.J., Dufour, A.P., McCabe, L.J. and Levin, M.A. (1982) Swimming associated gastroenteritis and water quality. *American Journal of Epidemiology 115(4)*, 606-616.

Cabelli, V.J., Dufour, A.P., McCabe, L.J. and Levin, M.A. (1983) A marine recreational water quality criterion consistent with indicator concepts and risk analysis. *Journal of the Water Pollution Control Federation 55*, 1306-1314.

Cabelli, V.J., Levin, M.A., Dufour, A.P. and McCabe, L.J. (1975) *The development of criteria for recreational waters.* In Gameson, A.L.H. (Ed.) *Discharge of sewage from sea outfalls.* Pergamon Press Oxford. 63-74pp.

Canadian Government (1983) *Guidelines for Canadian Recreational Water Quality.* Ministry of National Health and Welfare. Ottawa. 75p.

Cheung, W.H.S., Hung, R.P.S. Chang, K.C.K and Kleevens, J.W.L. (1990) Epidemiological study of beach water pollution and health-related bathing water standards in Hong Kong. *Water Science and Technology 23(1-3)*, 243-252.

Cheung, W.H.S., Kleevens, J.W.L., Chang, K.C.K. and Hung, R.P.S. (1988) Health effects of beach water pollution in Hong Kong. *Proceedings of the Institution of Water and Environmental Management.* Proceedings of the Annual Conference 376-383.

Colwell, R.R. (1987) Microbiological effects of ocean pollution. Paper 23 *International Conference on Environmental Protection of the North Sea.* 19p.

Denis, F.A., Blanchouin, E., Delignieres, A and Flamen, P. (1974) Coxsackie A16 infection from lake water. *Journal of the American Medical Association* 228, 1370-1371.

Denness, B. (1987) *Sewage disposal to the sea.* Sons of Neptune Ltd. Scarborough. 68p.

DoE (1989a) Department of the Environment *The quality of UK Bathing waters improves.* News Release 96 21.2.89. 19p. and Environmental Data Services (1989) Bulletin. *ENDS Report 177*, 3

DoE (1989b) Department of the Environment *Water Research Centre to carry out bathing water study.* Environment News Release No. 285 17th May. 2p.

DoE (1990) Environment News Release No 147. 5th March 1990. See also Parliamentary Question from Mr Douglas French (Con - Gloucester) to Mr Chris Patten.

Dufour, A.P. (1982) *Fresh recreational water quality and swimming associated illness.* Second National Symposium on Municipal Wastewater Disinfection, Orlando Florida. January 26-28, 21p.

Dufour, A.P. (1984) Bacterial indicators of recreational water quality. *Canadian Journal of Public Health* 75, 49-56.

EEC (1976) Council Directive of 8 December 1975 concerning the quality of bathing water (76/160/EEC). *Official Journal of the European Communities L/31*, 1-7.

EEC (1989) Proposal for a Council Directive concerning municipal waste water treatment COM(89) Final. Presented by the Commission, Brussels, 13th November 1989.

El Sharkawi, F. and Hassan, M.N.E.R. (1982) The relation between the state of pollution in Alexandria swimming beaches and the occurrence of typhoid among bathers. *Bull. High. Inst. Pub. Hlth. Alexandria* 12, 337-351.

Ellis, J.B. and Stanfield G. (1987) *Bacterial quality of marine bathing waters: a comparative assessment of North American and European standards.* Water Research Centre Report PRS 1595-M. 20p.

Fattal, B., Peleg-Olevsky, T. Agurshy and Shuval, H.I. (1986) The association between sea-water pollution as measured by bacterial indicators and morbidity of bathers at Mediterranean beaches in Israel. *Chemosphere 16(2-3)*, 565-570.

Ferley, J.P., Zimrou, D., Balducci, F., Baleux, B., Fera, P., Larbaigt, G. Jacq, E., Mossonnier, B. Blineau, A. and Boudot. J. (1989) Epidemiological significance of microbiological pollution criteria for river recreational waters. *International Journal of Epidemiology 18(1),* 198-205.

Fleisher, J.M. (1991) A reanalysis of data supporting US Federal bacteriological water quality criteria governing marine recreational waters. *Research Journal of the Water Pollution Control Federation 63(3),* 259-265.

Foulon, G., Maurin, J., Quoi, N.N. and Martin-Bouyer, G. (1983) Etude de la morbidite humaine en relation avec la pollution bacteriologique des eaux de baignade en mer. *Revue Francaise des Sciences de L'eau 2(2),* 127-143.

Galbraith, N.S., Barrett, N.J. and Stanwell-Smith, R. (1987) Water and disease after Croydon: a review of water-borne and water associated disease in the UK 1937-86. *Journal of the Institution of Water and Environmental Management 1(1),* 7-21.

Garrett, P. (1987) Bathing water - time for change. *Water Bulletin 284,* 6-7.

GESAMP (1990) IMO/ FAO/ UNESCO/ WAO/ WHO/ IAEA/ UN/ UNEP. Joint Group of Experts on the Scientific Aspects of Marine Pollution. *Report and Studies No.39 The State of the Marine Environment.* United Nations Environment Programme. 111p.

Grimes, J.D. (1986) *Assessment of ocean waste disposal. Task 5 Human health impacts of waste constituents II Pathogens and antibiotic - and heavy metal - resistant bacteria.* Final report submitted to the Congress of the United States Office of Technology Assessment Contract No 533-2685.0. 132p.

H.M.S.O. (1984) *Royal Commission on Environmental Pollution Tenth Report. Tackling Pollution-Experience and Prospects.* Cmnd. 9149 H.M.S.O. London. 233p.

H.M.S.O. (1985a) *House of Commons Committee on Welsh Affairs Coastal sewage pollution in Wales.* Minutes of Evidence 5.12.84. H.M.S.O. London. 117p. Minutes of Evidence 16.1.85 . H.M.S.O. London. 118-167pp.

H.M.S.O. (1985b) *House of Commons Committee on Welsh Affairs Coastal sewage pollution in Wales.* Report and Proceedings Vol I and II. 12.12.85. H.M.S.O. London. 27p.

H.M.S.O. (1985c) *House of Commons Committee on Welsh Affairs Coastal sewage pollution in Wales* Minutes of Evidence 5.12.84. H.M.S.O. London. 117p.

H.M.S.O. (1990a) *Pollution of Beaches.* House of Commons Environment Committee, Fourth Report, Volume I. H.M.S.O. London. lvii pp.

H.M.S.O. (1990b) *Pollution of Beaches.* House of Commons Environment Committee, Fourth Report, Volume II. H.M.S.O. London. 1-356 pp.

H.M.S.O. (1990c) *Pollution of Beaches.* House of Commons Environment Committee, Fourth Report, Volume III. H.M.S.O. London. 356-560 pp.

H.M.S.O. (1991) *Digest of water and environmental statistics No 13.* H.M.S.O.. London.

Havelaar, A.H. (1988) *F-Specific RNA Bacteriophages as model viruses in Water Treatment processes.* Rijksinstituut voor Volksgezondheid en Mileiuhygeine, Bilthoven. 239p.

Havelaar, A.H. and Olphen, M van. (1989) *Water quality standards for bacteriophages?* In Wheeler, D. et al. (Eds) *Watershed the future of water in Europe.* Pergamon, Oxford. 357-366pp.

Havelaar, A.H. and Pot-Hogeboom, W.M. (1988) F-specific RNA-bacteriophages as model viruses in water hygiene: ecological aspects. *Water Science and Technology 20(11/12)*, 399-407.

Hawley, H.B., Morin, D.P., Geraghty, M.E., Tomkow, J. and Phillips, C.A. (1973) Coxsackievirus B epidemic at a boy's summer camp. Isolation of virus from swimming water. *Journal of the American Medical Association 266*, 33-36.

Henderson, J.M. (1968) Enteric disease criteria for recreational waters. *Journal of the Sanitary Engineering Division 94*, 1253-1259.

Institute for European Environmental Policy. (1986) *Conclusions of the Montpellier Seminar.* Mimeographed.

Jones, F. (1981) Bacterial pollution of marine waters from the disposal of sewage sludge to sea. *Water Science and Technology 14*, 61-70.

Jones, F. and Kay, D. (1989) Bathing waters and health studies. *Water Services 93(1117)* 87-89.

Jones, F., Kay, D., Stanwell-Smith, R. and Wyer, M.D. (1990) An appraisal of the potential public health impacts of sewage disposal to UK coastal waters. *Journal of the Institution of Water and Environmental Management 4(3)*, 295-303.

Kay, D. and McDonald, A.T. (1986a) Bathing water quality: the significance of epidemiological research in the British context. *European Water and Sewage 90(1085)*, 321-328.

Kay, D and McDonald, A.T. (1986b) Coastal Bathing Water Quality. *Journal of Shoreline Management 2*, 259-283.

Kay, D., Wyer, M.D., McDonald, A.T. and Woods, N. (1990) The application of water quality standards to UK bathing waters. *Journal of the Institution of Water and Environmental Management 4(5)* 436-441.

Lightfoot, N.E. (1989) *A prospective study of swimming related illness at six freshwater beaches in Southern Ontario.* Unpublished PhD Thesis. 275p.

M.R.C. (1959) MEDICAL RESEARCH COUNCIL Sewage contamination of bathing beaches in England and Wales. *Medical Research Council Memorandum No. 37*, 32p.

Melnick, J.L. (1984) Enteric viruses in water. Monographs in Virology Vol 15. Karger. Basel. 235p.

Moore, B. (1975) *The case against microbial standards for bathing beaches.* In Gameson, A.L.H. (Ed.) *Discharge of sewage from sea outfalls.* Pergamon Press, London.

Moore, B. (1977) The EEC Bathing Water Directive. *Marine Pollution Bulletin 8(12)*, 269-272.

Mujeriego, R., Bravo, J.M. and Feliu, M.T. (1982) Recreation in coastal waters public health implications. *Vier Journee Etud. Pollutions, Cannes, Centre Internationale d'Exploration Scientifique de la Mer.* pp. 585-594.

National Academy of Sciences (1972) *Water Quality Criteria 1972.* A report of the Committee on Water Quality Criteria, Environmental Studies Board. National Academy of Sciences National Academy of Engineering, Washington DC. 594pp.

New Jersey Department of Health (1988) *A study of the relationship between illness in swimmers and ocean beach water quality.* State of New Jersey Health Department March 1988. 167p.

New Jersey Department of Health (1989) *A study of the relationship between illness in swimmers and ocean beach water quality interim summary report.* New Jersey Health Department March 1989. 54p.

NTAC (1968) National Technical Advisory Committee. *Water quality criteria.* Federal Water Pollution Control Administration, Department of the Interior, Washington DC.

NTAC National Technical Advisory Committee (1972) *Water Quality Criteria 1972.* A report of the Committee on Water Quality Criteria, Environmental Studies Board. National Academy of Sciences National Academy of Engineering, Washington DC. 594p.

PHLS (1959) PUBLIC HEALTH LABORATORY SERVICE. Sewage contamination of coastal bathing waters in England and Wales: a bacteriological and epidemiological study. *Journal of Hygiene, Cambs. 57(4),* 435-472.

Philipp, R., Waitkins, S., Caul, O., Roome, A., McMahon, S. and Enticott, R. (1989) Leptospiral and Hepatitis A antibodies amongst windsurfers and waterskiers in Bristol City docks. *Public Health 103,* 123-129.

Pike, E.B. (1990) *Health Effects of Sea Bathing (ET 9511SLG) Phase I Pilot studies at Langland Bay 1989.* Report DoE 2518-M. Water Research Centre, Medmenham, 109pp + 2 Appendices.

Seyfried, P.L., Tobin, R., Brown, N.E. and Ness, P.F. (1985) A prospective study of swimming related illness. I Swimming associated health risk. II Morbidity and the microbiological quality of water. *American Journal of Public Health 75(9),* 1068-1075.

Seyfried, P.L., Tobin, R., Brown, N.E. and Ness, P.F. (1985b) A prospective study of swimming related illness. II Morbidity and the microbiological quality of water. *American Journal of Public Health 75(9),* 1071-1075.

Shuval, H.I. (1986) *Thalassogenic diseases.* UNEP Regional Seas Reports and Studies No. 79. Published by the United Nations Environment Programme, Athens.

Snowdon, J.A. and Cliver, D.O. (1989) Coliphages as indicators of human enteric viruses in groundwater. *Critical Reviews in Environmental Control 19(3),* 231-249.

Stevenson, A.H. (1953) Studies of bathing water quality and health. *American Journal of Public Health 43,* 529-538.

Tyler, J. (1985) Occurrence in water of viruses of public health significance. *Journal of Applied Bacteriology Vol 59.* Symposium Supplement, 37S-46S.

USEPA. (1986) UNITED STATES ENVIRONMENTAL PROTECTION AGENCY. *Ambient water quality criteria for bacteria - 1986.* EPA440/5-84-002. Office of Water Regulations and Standards Division. Washington DC 20460. 18pp.

West, P.A. (1989) The human pathogenic vibrios - A public health update with environmental perspectives. *Epidemiology and Infection 103*, 1-34.

Wheeler, D. (1990) On the Beach. *Laboratory Practice 39(4)*, 19-24

WHO/UNEP (1990) *Microbiological pollution of the Mediterranean Sea. Long-term programme for pollution monitoring and research in the Mediterranean Sea (MED POL Phase II)* Report on a Joint WHO/UNEP Working Meeting. Valletta 13-16th December 1989. EUR/HFA Target 20. Published by the World Health Organisation Regional Office for Europe, Copenhagen. 17p.

WHO (1972) *Health criteria for the quality of recreational waters with special reference to coastal waters and beaches.* Ostend Belgium. 13-17th March. Published by the World Health Organisation Regional Office for Europe, Copenhagen 26p.

WRc (1989) Water Research Centre *Press Release*. 18th August 1989.

WRc (1990) Water Research Centre *Health Effects of sea bathing - Pilot study at Langland Bay, Swansea* 1989. Press Release 2p.

Chapter 11

TERMINAL DISINFECTION OF WASTEWATER WITH CONTINUOUS MICROFILTRATION

Vincent P. Olivieri and George A. Willinghan III
Memtec America Inc., Timonium, Maryland USA

INTRODUCTION

The terminal disinfection of wastewater effluents has been employed in some parts of the world to prevent the transmission of disease by the water route. Diseases that follow the faecal-oral route of transmission may be transmitted by direct consumption of the water, consumption of shellfish taken from the water and by close contact recreation in the water. Disinfection, as the word implies, is the removal or inactivation of infectious microorganisms such that disease is not detected. Disinfection is not sterilization of the water such that all living material is destroyed, but simply a reduction in the number of infectious agents such that disease is not spread by the use of the water after discharge to the environment.

Conventional wastewater treatment processes, including secondary treatment, effectively remove protozoan cysts but have little effect on the levels of bacteria and virus in the effluent unless terminal disinfection is practiced. Figure 11.1 shows the logarithm of the level of *Salmonella* spp. found in secondary effluent at three wastewater treatment plants over the course of one year in the United States (Sarai et al., 1989). *Salmonella* spp. were consistently recovered from the effluent of the three plants at levels of a few/gal to $1000 \cdot 100ml^{-1}$. In the absence of disinfection, these pathogens would be discharged to the environment. Figure 11.2 shows the level of male-specific bacterial virus in the effluent before disinfection at two wastewater treatment plants in the United States. The male-specific bacterial virus are consistently recovered from the effluent at relatively high levels. Both the bacterial and viral pathogens once discharged to the aquatic environment tend to die-away rather than proliferate. The rate at which the pathogenic bacteria die-away is

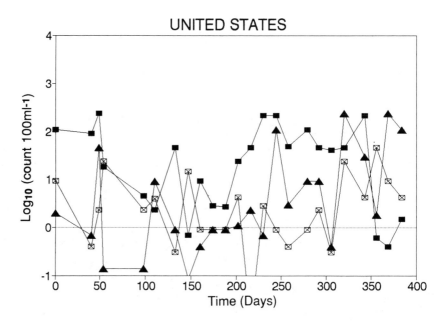

Figure 11.1 – Log levels of natural populations of salmonellae in undisinfected wastewater effluent at three plants in the United States.

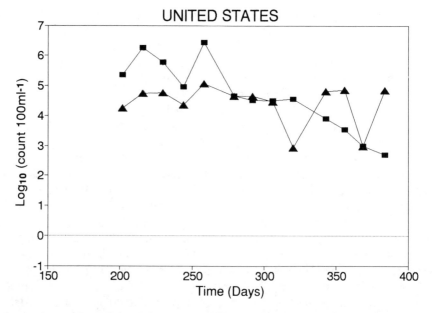

Figure 11.2 – Log levels of male-specific bacterial virus (host *Salmonella typhimurium* WG49) in undisinfected wastewater effluent at two plants in the United States.

dependent on many environmental factors but is slow enough for the disease causing microorganisms to survive in sufficient numbers to impact on drinking water supplies, shellfish harvesting areas and close contact recreational waters.

This chapter will present and discuss pilot and full scale information on the Memtec CMF and processes that incorporate CMF to treat wastewater from Australia, the United States and the United Kingdom. The quality of the CMF filtrate far exceeds any regulatory requirements for indicator bacteria. The CMF also removes virus, and reduces suspended solids (SS) and biochemical oxygen demand (BOD) to yield a consistently high quality advanced secondary effluent that could safely be discharged to the aquatic environment including bathing and recreational waters.

Until recently, the only means of providing for the disinfection of wastewater was to add strong reactive chemicals or the difficult to control application of ultra-violet light to inactivate the bacteria and viruses and reduce the numbers of these microorganisms in the effluent before discharge. The removal of microorganisms by membrane filtration has been common in the laboratory and in low volume applications for some time. Recent developments in membranes, membrane configuration and backwash technology by Memtec Limited has produced a continuous microfiltration (CMF) system capable of disinfecting wastewater and to dramatically improve the effluent quality without the addition of conventional harsh chemical disinfectants (Kolega et al., 1989 and Kolega et al., 1990). The Memtec CMF is a 0.2 micrometre polypropylene hollow fibre filter that automatically backwashes with air at a preset time interval or pressure drop. The Memtec CMF is operated in the direct flow or *dead end* mode at low pressure (30 to 70 Kpa) and produces a flux of about 100 to 150 $l.hr^{-1}$ square metre^{-1}) depending on inlet pressure and water quality. CMF can be applied as a final polishing process to disinfect and improve the effluent quality. Numerous, old trickling filter plants can be found around the world that are still functional but provide an effluent of insufficient quality to meet modern requirements.

RESULTS WORLD-WIDE

Memtec Continuous Microfiltration (CMF)
The village of Blackheath in the Blue Mountains of Australia discharges a secondary trickling filter effluent to a stream in the Blue Mountain National Park. The 4 Ml plant operated by the Sydney Water Board has installed Memtec CMF after conventional trickling filtration to demonstrate the applicability of the CMF technology. Figure 11.3 shows a schematic of the Blackheath CMF application. The settled, screened secondary effluent is pumped to an elevated tank to provide flow and pressure control. The trans-membrane pressure (TMP) for the system is about 60 Kpa. The effluent flows by gravity from the head tank to 8 blocks of 60, 2 m^2 filter area modules. The 120 m^2 blocks are operated independently and each is backwashed in turn, while the other blocks are in the filtration mode. The system is modular and provisions have been incorporated in the design for the addition of 2, 120 m^2 blocks

AUSTRALIA

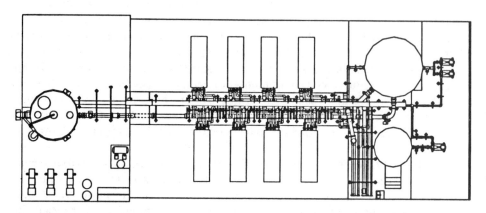

Figure 11.3 – Plan view of the 4 megalitre CMF at Blackheath in Australia.

as required. The system is land and civil works conservative and occupies a cement pad of 8 by 12 metres.

Table 11.1 shows data on the level of SS, BOD and faecal coliform before and after CMF obtained during the commissioning of the plant in the fall of 1990. Levels of suspended solids were generally less than 1 mg.l^{-1} with a maximum of 5 mg.l^{-1} and BOD was below the detection limit of 2 mg.l^{-1}. Faecal coliform were also at the detection limit and not recovered in the effluent. Data is continuing to accumulate on the efficacy of Memtec CMF under full scale conditions at this plant.

Long term pilot studies were conducted at the Round Corner treatment plant in New South Wales, Australia (Kolega et al., 1989 and Kolega et al., 1990). Round

Table 11.1 – The level of suspended solids (SS), biochemical oxygen demand (BOD) and faecal coliform before and after CMF obtained during the commissioning in the fall of 1990 of the 4 megalitre plant at Blackheath, New South Wales, Australia.

SAMPLE	SECONDARY EFFLUENT FEED			CMF FILTRATE		
	SS MG/L	BOD MG/L	FC, ESTIMATE #/100 ML	SS MG/L	BOD MG/L	FC #/100 ML
1	23	9	>1,000,000	<1	<2	<1
1	43	12	>1,000,000	5	<2	<1
3	24	9	>1,000,000	<1	<2	<1
4	22	6	>1,000,000	<1	<2	<1
MEAN	28	9	>1,000,000	<2	<2	<1

Corner is an extended aeration plant that treats domestic wastewater. The secondary effluent was fed to the Memtec CMF for filtration. Figure 11.4 shows the levels of SS (11.4a) and BOD (11.4b) before and after CMF. The SS and BOD varied considerably over the period of the study but, the CMF effluent was consistently below 2 mg.l^{-1} SS and BOD. Figure 11.5a shows the level of total coliform and faecal streptococci plotted as the logarithm of the count .100ml^{-1} before and after CMF. The level of the total coliform and faecal streptococci were quite variable with levels about $\log_{10}$ 6 and $\log_{10}$ 4 .100ml^{-1} (i.e. 1,000,000 .100ml^{-1} and 10,000 .100ml^{-1}), respectively. The levels of the indicator bacteria were below the detection limit in the CMF filtrate, less than 1 .100ml^{-1}. The 0 on the ordinate in Figure 11.5a was the logarithm of 1. Figure 11.5b shows the level of human enterovirus before and after CMF. Samples were concentrated and inoculated on monkey kidney, Buffalo Green Monkey (BGM), HEp-2 and human embryonic fibroblast (HEF) cell cultures. Quantal assays to determine TCID50 were performed with BGM cell cultures. Levels of human enterovirus were between $\log_{10}$ 2 (100) and $\log_{10}$ 5 (100,000) .50l^{-1} in the wastewater secondary effluent. Human enterovirus were not detected in the CMF filtrate during normal filter operation. $\log_{10}$ 0 represented the lower sensitivity limit of 1 TCID50 .50l^{-1}.

Similar pilot studies are being conducted in the United States. A pilot plant has been operating at the Back River Wastewater Treatment Plant in Baltimore, Maryland to demonstrate long term continuous application of CMF in the United States. Back River is a large activated sludge plant with activated sludge processes built at various times over the past 50 years. The Memtec CMF pilot plant uses effluent from the old activated sludge process. Figure 11.6 shows data collected remotely for 1 cycle of filtration. The pilot plant, consisting of 15 x 2 or 30 l. h^{-1} m^{-2} of filter area operates at a TMP of 30 to 70 KPa and a flux of about 120 of filter area (about 15-20 gals.min^{-1} [i.e. 68-91 l.min^{-1}]). Figure 11.7a and 11.7b show the levels of SS and BOD before and after CMF. Note, that the levels of SS and BOD vary considerably over the four months of operation in the activated sludge effluent (ASE). Despite the variation in the feed quality the CMF filtrate consistently provided an effluent of less than 2 mgl^{-1} of SS and BOD. Figure 11.8 shows the level of indicator bacteria (11.8a) and male-specific bacterial virus (11.8b) as the logarithm of the number .100ml^{-1} before and after CMF. The levels of total coliform, faecal coliform and enterococci are high in the ASE. With the exception of the initial total coliform sample run in the new Memtec America Water and Wastewater laboratory, the CMF filtrate contained levels below the detection limit of 1 .100ml^{-1} (logarithm=0) for the indicator bacteria. Male-specific bacterial virus, proposed by some (Havelaar et al., 1984; Grabow et al., 1984; Havelaar et al., 1989 and Grabow et al., 1989) as surrogates for human virus, were also removed to the detection limit at the Back River CMF pilot plant.

MEMTEC MEMBIO™ BIOLOGICAL AERATED FILTER+CMF

The CMF can also be applied in conjunction with a biological aerated filter to provide a series of processes to treat raw sewage. The MEMBIO™ process takes raw

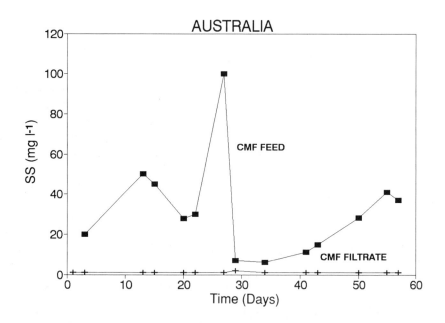

Figure 11.4a –Levels of suspended solids (SS) before and after CMF filtration in extended aeration effluent at the Round Corner pilot plant in Australia.

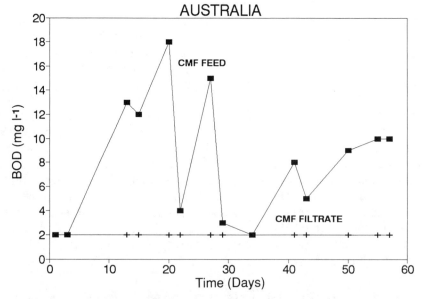

Figure 11.4b –Levels of biochemical oxygen demand (BOD) before and after CMF filtration of extended aeration effluent at the Round Corner pilot plant in Australia.

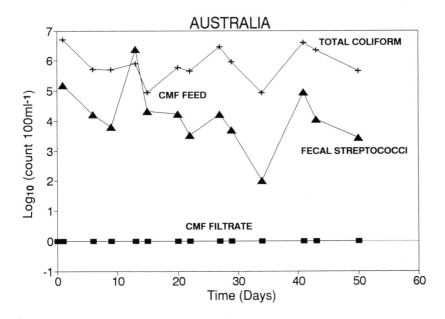

Figure 11.5a –Log levels of indicator bacteria before and after CMF filtration of extended aeration effluent at the Round Corner pilot plant in Australia.

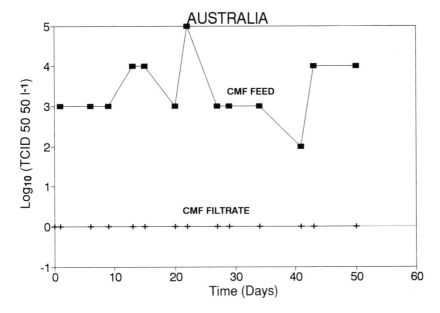

Figure 11.5b –Log levels of human enterovirus before and after CMF filtration of extended aeration effluent at the Round Corner pilot plant in Australia.

wastewater after preliminary treatment (screening and grit removal) into a fixed film submerged reactor. Air is continually supplied to the reactor. Periodically a pulse of air is put to the reactor in a proprietary process to minimize channeling and bed imperfections. The bioreactor is backwashed about once a day and the backwash is collected for subsequent thickening and digestion by conventional anaerobic digestion. The sludge generated resembles the humus type generated in secondary sedimentation by trickling filter plants. The effluent from the bioreactor is clarified by CMF. The MEMBIO™ process provides for rapid return of still active biomass to the bioreactor. Active biosorption occurs in the bioreactor and at the barrier layer on the membrane at the microfilter. The MEMBIO™ process requires short contact times in the bioreactor for biosorption by the active mass of the biofilm and the separation of biological solids from the effluent is accomplished rapidly by CMF. The process is, thus, compact and requires only a fraction of the land needed by conventional processes.

The MEMBIO™ process produces a high quality effluent. Figure 11.9a shows the removal of SS and 11.9b shows the removal of BOD by the MEMBIO™ process. The feed varied from 100 to more than 200 mgl^{-1} for SS and BOD during pilot trials at Malabar in Australia. After the bioreactor plus the CMF (MEMBIO™) the effluent contained a consistently low levels of SS and BOD with a mean SS of about 2 mgl^{-1} and mean BOD of about 7 mgl^{-1}. Microbial levels shown in Figure 11.10 are similar to those observed for CMF alone. Fecal coliform were not found in the MEMBIO™ effluent. The MEMBIO™ performed so well in pilot that a full scale 40 Ml day^{-1} plant is being built at Crunulla in New South Wales, Australia.

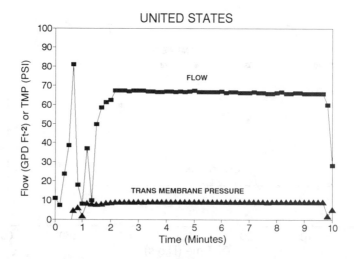

Figure 11.6 – CMF flow (gallons per day per square foot of membrane area) and trans membrane pressure (TMP) for one filter per cycle for dead end operation at the Back River WWTP in the United States.

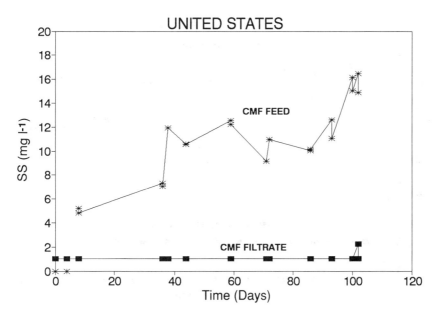

Figure 11.7a –Levels of suspended solids (SS) before and after CMF filtration of activated sludge effluent at the Back River pilot plant in the United States.

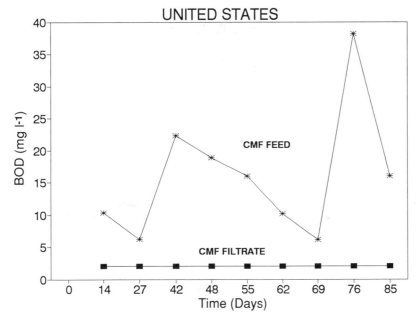

Figure 11.7b –Levels of biochemical oxygen demand (BOD) before and after CMF filtration of activated sludge effluent at the Back River pilot plant in the United States.

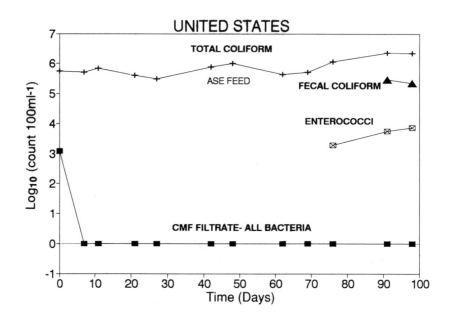

Figure 11.8a – Log levels of indicator bacteria before and after CMF filtration of activated sludge effluent at the Back River pilot plant in the United States.

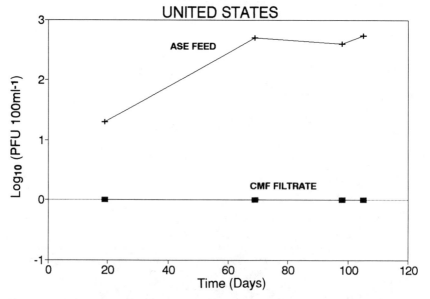

Figure 11.8b – Log levels of male-specific bacterial virus before and after CMF filtration of activated sludge effluent at the Back River pilot plant in the United States.

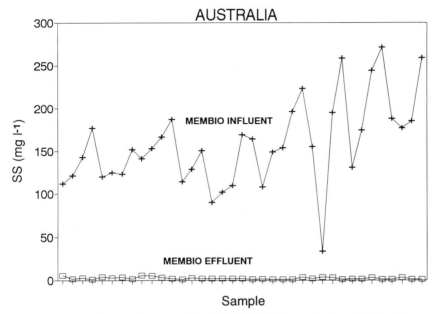

Figure 11.9a –Levels of suspended solids (SS) before and after the Membio process during trials at Malabar in Australia.

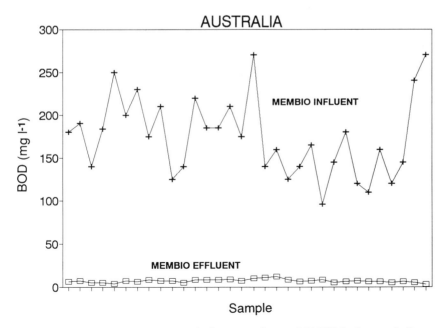

Figure 11.9b –Levels of biochemical oxygen demand (BOD) before and after the Membio process during trials at Malabar in Australia.

The results of initial MEMBIO™ pilot trials in the United Kingdom are shown in Figures 11.11 and 11.12. The SS and BOD before and after MEMBIO™ are shown in Figure 11.11a and 11.11b, respectively. The input SS and BOD was more variable than in previous studies. The SS varied from 40 to greater than 500 mgl^{-1} and the BOD varied from just less than 50 to greater than 400 mgl^{-1}. The MEMBIO™ process produced a high quality effluent of less than 20 mgl^{-1} SS and 30 mgl^{-1} BOD. The levels of faecal coliform and faecal streptococci in the MEMBIO™ effluent were less than 10 .100ml^{-1}, the detection limit for the assay conducted (Figure 11.12a). The levels of different groups of bacterial virus were also low in the filtrate. Male specific bacterial virus (Havelaar host WG49, Havelaar *et al.*, 1984) were not found in the MEMBIO™ effluent. MS$_2$ virus was found in 2 of 10 samples at 4 and 1 plaque forming units (PFU)ml^{-1} where the input was 3400 and 2200 PFUml^{-1}, respectively. Greater than 99.9% virus removal was observed.

DISCUSSION

One of the basic concepts in the field of public health is to contain the faeces and institute a system of multiple barriers to prevent the spread of pathogenic microorganisms found in the faeces of morbid to healthy individuals. In wastewater treatment, the containment is accomplished by the collection of human wastes in sewers and transmission to a central location for treatment. Conventional wastewater treatment without disinfection accomplishes little in terms of removal of indicator and pathogenic microorganisms. Table 11.2 (Olivieri *et al.*, 1989; Olivieri *et al.* 1990) shows the relative microbial balance for selected indicator and pathogenic microorganisms for three typical treatment plants in the United States. The relative mean log microbial mass was calculated by multiplying the mean log microbial level

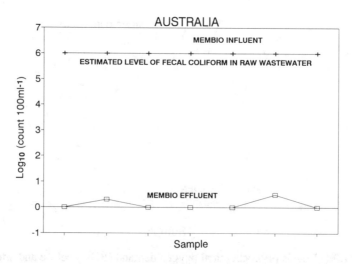

Figure 11.10 – Log levels of faecal coliform before and after the Membio process during trials at Malabar in Australia.

Table 11.2 – Relative microbial balance from raw wastewater for the selected microorganisms at three plants in the United States. TC=total coliform, FC=faecal coliform, FS=faecal streptococci, ENT=enterococci, C.p.=*Clostridium perfringens*, Sal.=*Salmonella* spp., MSBV=male-specfic bacterial virus and HEV=human enterovirus.

	TC	FC	FS	ENT	C.p.	Sal.	MSBV	HEV
PLANT A		RELATIVE MEAN LOG MICROBIAL MASS						
RAW	10.29	9.62	8.05	7.87	6.15	4.37	8.02	3.30
DIGESTER IN	9.15	8.33	6.64	6.48	5.46	3.30	5.98	1.85
DIGESTER OUT	7.90	6.95	5.47	5.38	5.06	1.42	5.04	0.75
EFFLUENT	9.27	8.64	7.08	6.77	5.97	3.52	6.88	3.22
		MEAN LOG REDUCTION FROM RAW SEWAGE						
DIGESTED	2.40	2.68	2.58	2.49	1.08	2.95	2.98	2.55
EFFLUENT	1.02	0.98	0.98	1.10	0.18	0.85	1.13	0.07
PLANT B		RELATIVE MEAN LOG MICROBIAL MASS						
RAW SEWAGE	8.88	7.83	6.73	6.36	4.66	2.44	6.16	1.80
DIGESTER IN	8.46	7.33	6.05	5.74	4.43	2.10	5.31	0.95
DIGESTER OUT	5.49	4.51	4.01	3.61	3.61	-1.23	2.29	-0.42
EFFLUENT	6.97	5.84	4.93	4.60	4.07	1.34	5.26	1.81
		MEAN LOG REDUCTION FROM RAW SEWAGE						
DIGESTED	3.40	3.32	2.72	2.76	1.05	3.67	3.87	2.23
EFFLUENT	1.92	1.99	1.81	1.77	0.59	1.10	0.90	-0.00
PLANT C		RELATIVE MEAN LOG MICROBIAL MASS						
RAW SEWAGE	10.37	9.39	8.36	8.07	6.25	4.42	8.04	3.51
DIGESTER IN	9.29	8.33	6.59	6.48	5.63	3.52	6.03	2.04
DIGESTER OUT	7.75	6.98	5.89	5.96	5.94	2.10	4.69	1.59
EFFLUENT	7.51	6.78	5.68	5.78	5.77	2.80	5.18	3.48
		MEAN LOG REDUCTION FROM RAW SEWAGE						
DIGESTED	2.62	2.40	2.47	2.11	0.31	2.32	3.35	1.92
EFFLUENT	2.87	2.60	2.68	2.29	0.48	1.62	2.85	0.03

CALCULATION BASED ON THE MICROBIAL LEVELS AS NO.100 ml^{-1} AND THE MEAN FLOW IN MILLION GALLONS/DAY

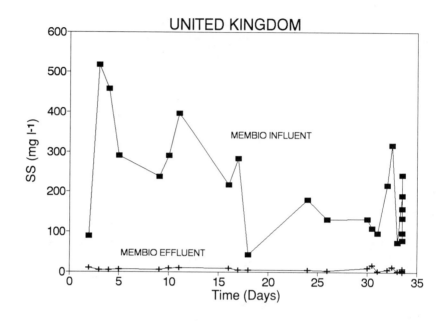

Figure 11.11a –Levels of suspended solids (SS) before and after the Membio process at a pilot plant in the United Kingdom.

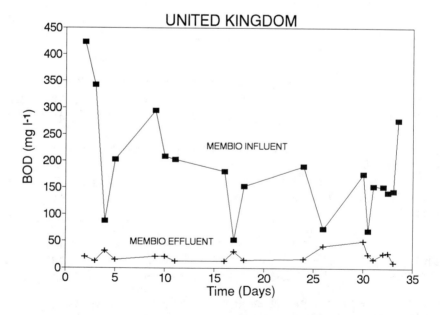

Figure 11.11b –Levels of biochemical oxygen demand (BOD) before and after the Membio process at a pilot plant in the United Kingdom.

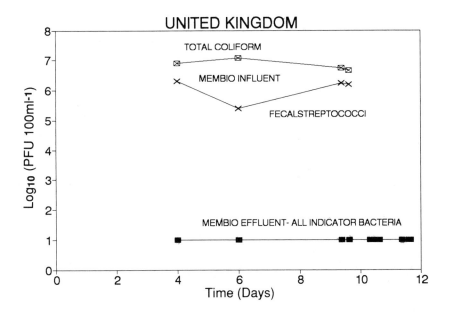

Figure 11.12a –Log level of indicator bacteria before and after the Membio process at a pilot plant in the United Kingdom.

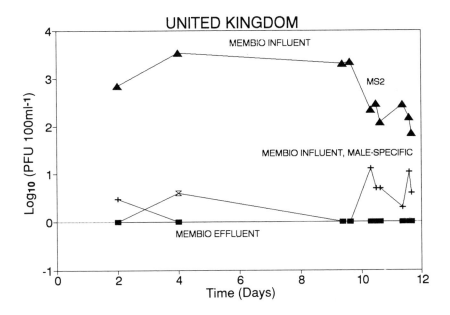

Figure 11.12b –Log level of bacterial virus before and after the Membio process at a pilot plant in the United Kingdom.

over approximately one year (in number .100ml^{-1}) times the average flow for the waste stream, in MGD. The overwhelming relative mass of microorganisms leaves the wastewater treatment plant in the wastewater effluent. More than 10 times the total coliform, faecal streptococci, enterococci, *Salmonella* spp. and male specific bacterial virus leave the plant in the liquid effluent than in the digested sludge (anaerobic). The mean log reduction through conventional secondary treatment processes is less than 90%. Without terminal disinfection of wastewater, a large number of disease causing microorganisms are returned to the aquatic environment where the potential exists for exposure to humans that ingest the water either by direct consumption, bathing, close contact recreation or consumption of shellfish harvested from the contaminated water. The initial containment and multiple barriers are no longer in place.

CONCLUSIONS

Membrane technology exists to remove mechanically microorganism from wastewater to protect down-stream bathing and recreational waters, water supplies and shellfish harvesting waters. This technology exists to disinfect wastewater with out the use of strong reactive chemicals which may produce toxic and/or mutagenic compounds (see Cairns, this volume). Full scale and pilot systems have been operated around the world to demonstrate the reliability and performance of Memtec CMF. The Memtec CMF either as a final polishing step or as a complete process coupled with a bioreactor (MEMBIO™) yields a consistent high quality effluent with a reliable high degree of disinfection for safe discharge to the aquatic environment.

ACKNOWLEDGEMENTS

The authors gratefully acknowledge the Sydney Water Board for providing the sites and facilities in Australia. The authors also wish to thank the Baltimore Bureau of Wastewater and Mr. Gerald Slatery and Mr. Gary Wagner at the Back River Wastewater Treatment for their kind assistance in siting the Memtec pilot facility in the United States.

REFERENCES

Grabow, W.O.K., Coubrough, P., Nupen, E.M. and Bateman, B.W. (1984). Evaluation of coliphages as indicators of the virological quality of sewage-polluted water. *Water South Africa.* 10(1), 7-14.

Grabow, W.O.K., Idema, G.K., Coubrough, P. and Bateman, B.W. (1989). Selection of indicator systems for human viruses in polluted seawater and shellfish. *Water Science Technology.* 21(3), 111-117.

Havelaar, A.H., Hogeboom, W.M. and Pot, R. (1984). F-Specific RNA bacteriophages in sewage: methodology and occurrence. *Water Science and Technology, 17,* 645-655.

Havelaar, A.H. and Pot-Hogeboom, W.M. (1988). F-specific RNA bacteriophages as model viruses in water hygiene: ecological aspects. *Water Science and Technology. 20 (11/12),* 399-407.

Kolega, M., Kaye, R.B., Chiew, R.F., and Grohmann, G.S. (1989). Disinfection of secondary sewage effluent by advanced membrane treatment technology. *Proceedings of Australian Water and Wastewater Association, 13th Federal Convention. 2,* pp 623-627.

Kolega, M., Grohmann, G.S., Chiew, R.F., and Day, A.W. (1990). Disinfection and Clarification of Treated Sewage by Advanced Microfiltration. *Proceedings of the International Association on Water Pollution Research and Control - 25th Anniversary Conference and Exhibition - Kyoto, Japan, July 29-August 3, 1990.*

Olivieri, V.P., Cox, L., Sarai, M., Sykora, J.L., and Gavaghan, P. (1990). Levels and Removal of Selected Indicator and Pathogenic Microorganisms During Conventional Anaerobic Sludge Digestion. *Proceedings of the Twelfth United States-Japan Conference on Sewage Treatment Technology* USEPA.

Olivieri, V.P., Cox, L., Sarai, M., Sykora, J.L., and Gavaghan, P. (1989). Reduction of Selected Indicator and Pathogenic Microorganisms During Conventional Anaerobic Sludge Digestion. *Proceedings of the AWWA/WPCF Joint Residuals Management Conference.* Water Pollution Control Federation, Alexandria, Virginia.

Sarai, M., Olivieri, V.P., Cox, L., Sykora, J.L., and Gavaghan, P. (1989). Relationship Between Levels of Selected Indicator and Pathogenic Microorganisms in Sewage and Sludge Samples From Three Sewage Treatment Plants. *Proceedings of the AWWA/WPCF Joint Residuals Management Conference.* Water Pollution Control Federation, Alexandria, Virginia.

Chapter 12

UV DISINFECTION : AN OVERVIEW OF PROCESSES AND APPLICATIONS

William L. Cairns
Research Manager, Trojan Technologies Inc., 845 Consortium Court, London, Ontario, Canada N6E 2S8

THE PROCESS OF UV DISINFECTION

PHOTOCHEMICAL CHANGES TO NUCLEIC ACIDS WITHIN MICROBES, THE IMPLICATIONS FOR APPLICATION

Ultraviolet (UV) light is an effective biocidal agent, whether the UV light comes from solar radiation penetrating the atmosphere (Bailey *et al.*, 1983) or from UV emitting lamps engineered into a disinfection system. Irradiation with UV is described as a physical disinfection method because only light (a physical agent) and no chemicals are used in the process. However, absorption of the UV light by certain molecules within the exposed microbes does bring about intracellular chemical changes which can prevent the microbes from reproducing or otherwise functioning. Figure 12.1 portrays the portion of the electromagnet spectrum which includes UV. Ultraviolet light is at a shorter wavelength (more energetic) region of the spectrum than visible light. The units used to measure UV wavelengths are nanometres (nm). UV is often classified as UVC (less than 280nm), UVA (greater than 315nm), and UVB between these two regions. Some literature will refer to *near UV* (300-380nm) and *far UV* (<300nm).

Absorption of UV light by a molecule results in excitation of the bonding and non-bonding electrons in the molecule to create more energetic states which can undergo photochemical reactions leading to new molecular structures. When these new structures interfere with the biological function of the transformed microbial molecule, then the microbe has only a limited number of options; (i) to rely upon additional copies of the transformed molecule to perform the critical functions, (ii) to replace the transformed molecule if it can, (iii) to repair the transformed molecule

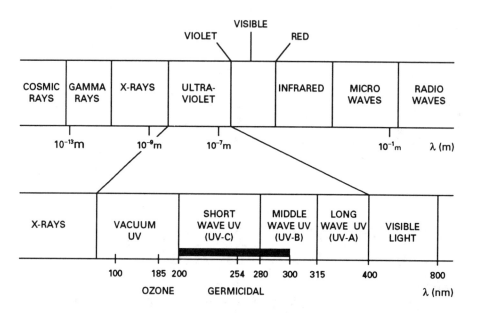

Figure 12.1 – Electromagnetic Spectrum with Expanded Scale of Ultraviolet Radiation. (1 nanometre = 10^{-9} metre).

if it can, or (iv) to cease to function as a viable cell when the biological system has sufficiently deteriorated due to an absence of functional critical molecules.

When irradiated with UV below 300nm (UVC and part of UVB), the target UV-absorbing molecules are the nucleic acids which are polymers encoding the genetic information of the cell. That the nucleic acids (DNA for bacteria and most viruses, but RNA for some viruses) are in fact the target molecules whose transformation results in inactivation of microbes is illustrated by parallels in the wavelength dependence for microbial inactivation by UV and the absorption spectrum of nucleic acids (Figure 12.2).

Nucleic acids are especially critical components of the cell. Except for a short interval of time during the cell cycle, there is only one copy of the replicating nucleic acid within the cell. If this sole copy is transformed functionally by UV, then the microbe can have difficulty in addressing its own genetic information during microbe reproduction (which requires nucleic acid replication) or during repair of the damaged nucleic acid (by dark repair mechanisms involving excision of the damaged region and resynthesis using the complementary strand as template). Even when the microbe has an unique repair mechanism for certain kinds of photochemical transformations of the nucleic acids (i.e. a light dependent repair mechanism called photoreactivation) the presence of still other kinds of transformations may still leave a non-replicatable and otherwise unrepairable nucleic acid. The microbe is then without a means of reproducing itself, or of replacing enzymatic or structural proteins which undergo a normal turnover within cells or which must be newly synthesized at different physiological stages of the microbial cell cycle.

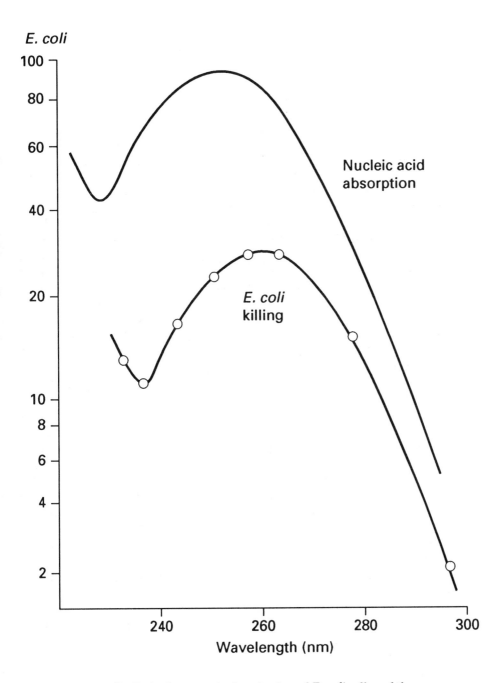

Figure 12.2 – Similarity between the inactivation of *E. coli* cells and the absorption spectrum of nucleic acids. (from Harm, W. "Biological Effects of Ultraviolet Radiation" Pg. 29, Cambridge University Press, Cambridge, 1980).

Absorption of UV by nucleic acids results in formation of various photoproducts, some of which are shown in Figure 12.3, and all of which can contribute to lethality. The most frequent is the cyclobutadithymine dimer, thy<>thy, formed by covalent bonding between two adjacent thymine bases (one of the building blocks and genetic information encoding components of the nucleic acids). Other reaction products include nucleic acid crosslinks with proteins, and the pyrimidine(6-4)pyrimidone photoadduct, thy(6-4) pyo, which has attracted attention because it is produced less than one tenth as frequently as the thy<>thy dimer, but has been suggested to be more important for mutations than the thy<>thy dimer in the bacterium *E. coli*, although the cyclobutane dimers (perhaps those involving the base cytosine as well as thymine) are suggested to be more important for mutations in human cells (Ossanna *et al.*, 1987; Brash, 1988). Arguments for the dimer also being mutagenic in *E. coli* have been presented (Hutchinson, 1987).

Many of the properties of cells which influence their susceptibility to chemical disinfectants have less impact on UV disinfection. Microbes generally have a net negative charge which must be overcome by an approaching negatively charged disinfectant molecule such as hypochlorite ions. UV encounters no such barrier. Carbohydrate exopolymeric material (slimes) which surround microbial cells depend on the microbe and its nutritionally dependent physiological state. In addition, cell walls composed of carbohydrate and non-aromatic amino acids, present diffusion barriers to chemical disinfectants. These can react slowly with chlorine to consume it on its way to more critical cell membrane and intracellular sites. These same carbohydrates and peptides of slimes and cell walls do not appreciably absorb UV in competition with nucleic acids. The ability of microbial cell surfaces to act as a diffusion barrier to chemical disinfectants but not a barrier to UV, may account for the observation (Yip and Konasewick, 1972) that the dose requirement to inactivate different microbes with UV shows less variation than with chemical disinfectants. Figure 12.4 shows this difference between UV and chemical disinfectants. An awareness of differences in the sensitivity of pathogens and indicator organisms and the target level of disinfection to specify for that indicator organism is necessary. Figure 12.4 would indicate that to achieve the same level of inactivation of the Polio and Coxsackie viruses with UV and chlorine, a lower level of indicator *E. coli* would have to be obtained with chlorine than UV. Other authors (Harris *et al.*, 1987; Chang *et al.*, 1985) have reported a 3-10 times higher UV dose being needed for viruses than for *E. coli*, but the dose covered a narrower dose range than the 25-43 times higher doses reported for chlorine disinfection of some viruses. These observations are important for epidemiological studies which attempt to find correlations between an incidence of disease and the presence of an indicator organism. The variables which can impact on finding such a correlation must be recognized:

(i) variation in the disease causing pathogens and their concentration from site to site but more importantly from time to time at a given site depends on which infectious pathogens are being delivered

Ch. 12   UV disinfection : an overview of processes and applications   179

Class of photoproduct	Representative photoproduct
Cyclobutane-type dimers	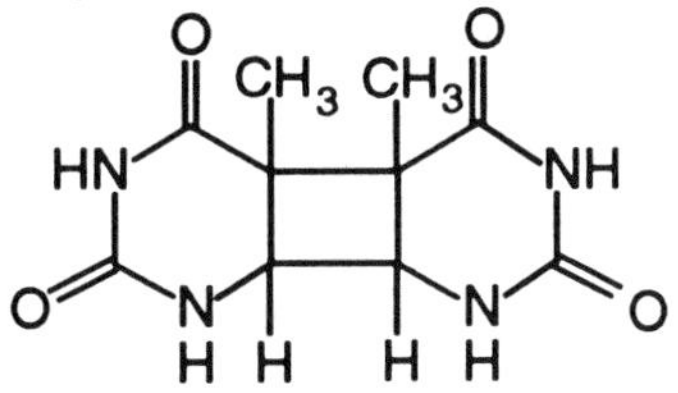 cys-syn thymine-thymine dimer
Pyrimidine adducts	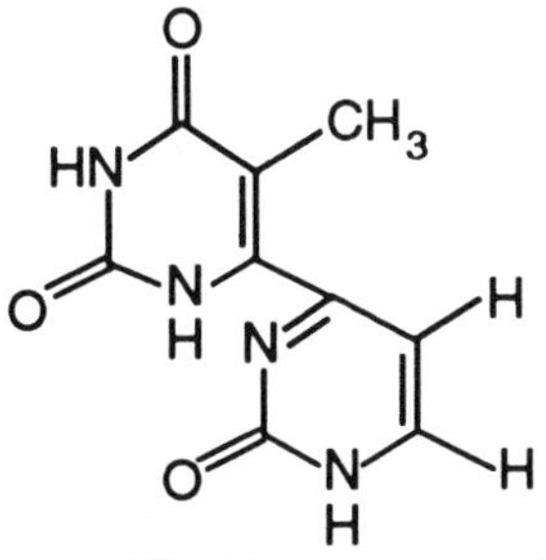 6-4'-[pyrimidin-2'-one] pyrimidine
Spore photoproduct	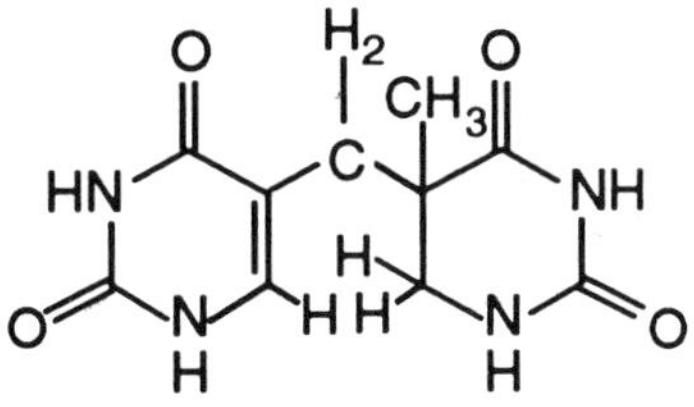 5-thyminyl-5,6-dihydrothymine
Pyrimidine hydrates	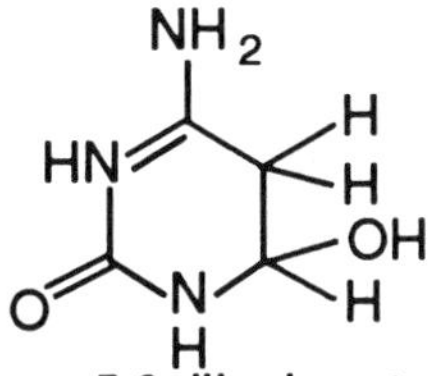 6-hydroxy-5,6-dihydrocytosine
DNA-protein crosslinks	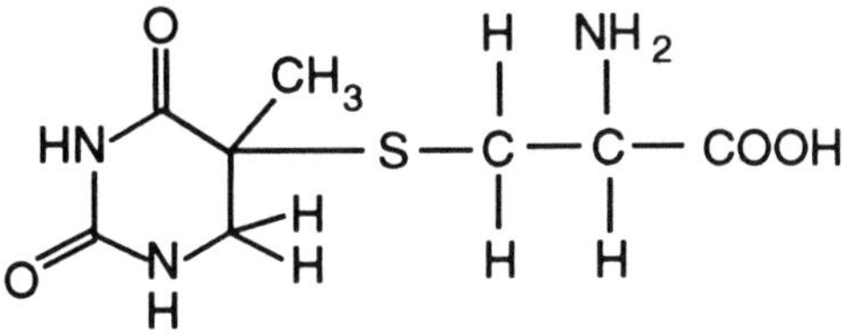 5-S-cysteine-5,6-dihydrothymine

Figure 12.3 – Examples of DNA photoproducts formed in UV-irradiated cells. (from Harm, W. "Biological Effects of Ultraviolet Radiation" Pg. 29, Cambridge University Press, Cambridge, 1980).

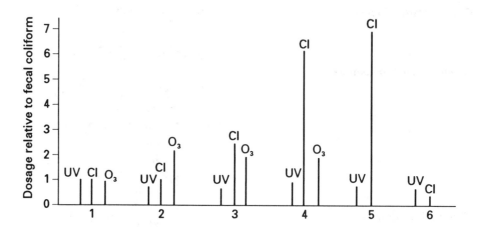

Figure 12.4 – Demonstration of relative effectiveness of UV:
(1) *Escherichia coli* B (2) *Salmonella typhosa*
(3) *Staphylococcus aureus* (4) Polio type 1 virus
(5) Coxsackie AZ virus (6) Adenovirus Type 3.

in infectious doses by the community through the sewage to the receiving water at the moment,

(ii)   the variation in susceptibility of the different pathogens and the indicator organism to different disinfection processes which may be used, and

(iii) differential survival rate of the high infectious dose pathogen of the moment and the indicator organism depending on stresses in the receiving water.

UV disinfection would appear to offer an advantage due to its lower dependence on microbial structure: if any appropriate universal indicator organism for human faecal contamination is used (i.e., an organism which indicates faecal contamination and does not die off faster than the pathogens, then monitoring the concentration of the indicator is likely to provide reasonable assurance of the control of the pathogens present, even if they vary from time to time or site to site. Table 12.1 shows the variation in UV dose (see below) required to inactivate microbes to the same level. The variation can be accounted for by some differences in UV absorbing components of the cell surface, differences in cell size where larger sized cells have more intracellular protein and RNA nucleic acids which filter out UV on its way to the DNA nucleic acids, differences in nucleic acid structures where the ratio of UV-absorbing component bases varies, and differences in the efficiencies of repair mechanisms.

Table 12.1 – Comparative sensitivity of microbes to UV disinfection.

MICROBE	DOSE (mW.s.cm^{-2}) FOR 90 REDUCTION IN COUNTS
BACTERIA	
Bacillus anthracis spores	54.5
Bacillus subtilus spores	12.0
Clostridium tetani	12.0
Corynebacterium diptheriae	3.4
Escherichia coli	3.4
Legionella pneumophila	1.0
Nicrococcus radiodurans	20.5
Mycobacterium tuberculosis	6.0
Pseudomonas aeruginosa	5.5
Salmonella enteritidis	4.0
Salmonella paratyphi	3.2
Salmonella typhi	2.1
Salmonella typhimurium	8.0
Shigella dysenteriae	2.2
Staphylococcus aureus	5.0
Streptococcus pyogenes	2.2
Vibrio comma	6.5
VIRUSES	
Bacteriophage	3.6
Influenza virus	3.6
Poliovirus	3.2
YEASTS	
Saccharomyces cerevisiae	7.3
MOULDS	
Penicillium roqueforti	14.5
Aspergillus niger	180.0
PROTOZOA	
various	60.0 to 200.0

NB - TYPICAL MINIMUM DOSE USED IN WASTEWATER DISINFECTION IS 30 mW.s.cm^{-2}

The thy<>thy dimer is repairable by the photoreactivating enzyme, but other dimers involving cytosine (such as thy<>cyt, cyt<>cyt), and the thy(6-4)pyo photoadduct can not be repaired except by the dark repair mechanisms. Photoreactivation is accomplished by an enzyme which can recognize the thy<>thy dimer in nucleic acids, bind to that site and upon absorption of a photon of longer wavelength light, split the dimer back to its original genetically functional structure (Hutchinson, 1987; Smith *et al.*, 1987). Photoreactivation of *E. coli* resulting from exposure of UV-irradiated cells to solar radiation will therefore reduce the lethality of UV irradiation, but not increase the number of mutants formed.

The practical significance of photoreactivation is difficult to assess, but field studies indicate it to be far less important than suggested by laboratory studies or model studies. Inactivation of *E. coli* requires several hundred transformations which must be repaired if the cell is to survive. The higher the number of transformations, the less the probability for complete repair. Repair could be either by photoreactivation or excision repair in the dark. Several studies (Harris *et al.*, 1987; Scheible and Bassel, 1981; Whitby, *et al.*, 1984) indicate that, under lab conditions, or when UV irradiated effluent is held within a jar suspended in the effluent channel, repair will occur. Jar suspended effluent can amount to 1 log reactivation over dark controls (Whitby, *et al.*, 1984), and laboratory studies can result in 3.4 and 2.4 log reactivation for *E. coli* and *Streptococcus faecalis* under ideal conditions (Harris *et al.*, 1987). Microbes released into a receiving water seldom face ideal conditions of light for photoreactivation, of nutrition, of freedom from predation by other microbes, of freedom from sedimentation or even of freedom from solar UV irradiation. In an important study (Palmateer and Whitby, 1987), a special strain of *E. coli* was irradiated with UV and/or non-irradiated before being released into the receiving waters. Figure 12.5 shows that both irradiated and non-irradiated microbes underwent a similar decay rate during their trip down the river. Any reactivation which occurred was minimal with respect to the overall decay rate.

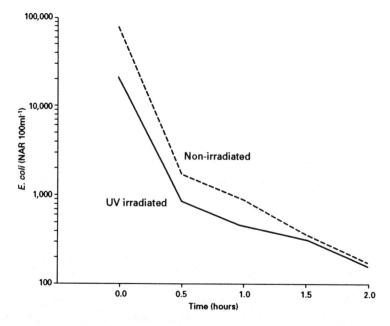

Figure 12.5 – Counts of UV irradiated and non-irradiated *E. coli* (NAR) in Big Otter Creek at different times after discharge to the receiving water.

A perspective on mutation is necessary to overcome apprehension about mutant super-pathogens being produced by UV disinfection. The spontaneous rate of mutation resulting from errors in copying genetic information during normal cell reproduction is about $1/10^6$, which for a typical raw sewage and primary effluent containing 10^6-10^7 faecal coliforms per 100ml results in about 1-10 mutants.100ml^{-1} per cell division (approximately every hour). Both solar UV and human-engineered UV could be used to reduce the secondary effluent levels of faecal coliforms from 10^4-10^6.100ml^{-1} down to 100.100ml^{-1}. The solar UV intensities would require hours to achieve this end, and the engineered UV intensities would take seconds. Under the worst case scenario, all 100 of the surviving faecal coliforms .100ml^{-1} from an engineered system using lamps emitting primarily 254 nm UV would be mutants. Solar UV would produces about 1/10 the number of mutants (Peak et al., 1984), i.e. about 10 mutant faecal coliforms .100ml^{-1}.

If these mutants are considered in terms of their numbers .m^{-3}, the total mutant faecal coliforms generated in this volume would be approximately as shown in Table 12.2 below:

Table 12.2 – Mutation rates for faecal coliform organisms.

Source	mutation process	mutant faecal coliforms .m^{-3}
primary effluent	spontaneous mutation	10^4 - 10^5
secondary effluent	spontaneous mutation	10^2 - 10^3
secondary effluent	solar UV	10^5
secondary effluent	254nm UV	10^6

This would appear like a large number of mutants .m^{-3} even though it may be no more in magnitude than the total number of faecal coliforms discharged in 10 ml (10^{-5} m^3) of a concentrated primary effluent. However, several further considerations are important:

(i) A mutation has no value to the microbe unless it brings about a selective functional advantage to the microbe. Some indication of the probability of this happening is indicated by early work showing the frequency with which a tryptophan amino acid requiring E. coli mutated to tryptophan independence (Peak et al., 1984). At doses of UV which reduced the population to 0.5% survivors, less than 1 in 1000 survivors had a mutated tryptophan synthesis functionality. UV irradiation of the *Salmonella typhimurium* TA98 strain used in the Ames mutagenesis tests produced a 10 fold lower level of such mutants (Calkins, *et al.*, 1987). Of the 10^6 faecal coliform mutants .m^{-3} released following treatment of secondary effluents with engineered UV systems, only 104 might be expected to have some select functional advantage which could have pathological or clinical significance. This is probably an upper value since the microbes of concern are already pathogens and mutations produced from them are more likely to interfere with the mechanisms of pathogenicity or to slightly enhance pathogenicity rather than to create pathogenicity where none existed before. Some secondary effects of a chemical or physical disinfection process may be selection of certain strains within a population due to interaction of the disinfectant with the microbes' chemical structure or physiological state as opposed to the disinfectant having created new strains through changes to the microbes' genetic structure. UV has been suggested (Meckes, 1982) to kill both antibiotic resistant and sensitive microbes but to increase amongst the survivors the percent of microbes which are resistant to some antibiotics (tetracycline and chloramphenicol), but not to other antibiotics (streptomycin). This does not appear to be due to mutation, but to, for example, the presence of a protein in the bacterial envelope which can provide both protection against the antibiotic and, by absorbing UV, also provide protection from UV reaching the nucleic acid target molecule. Chlorination of sewage effluents was reported also to increase the proportions of antibiotic-resistant organisms (Murray, *et al.*, 1984). In a separate study with microbes which were resistant and non-resistant to kasaugamycin and novobiocin, resistance provided no advantage for survival in the receiving waters, but three of the resistant strains died off more rapidly than non-resistant strains (Pettibone *et al.*, 1987). These changes in antibody resistance within microbial populations, as a consequence of disinfection, are as inevitable as the general increase in antibiotic resistance within the population as a result of using antibiotics in the practice of medicine.

(ii) The mutant must survive natural consuming phenomena in the receiving water: dilution, sedimentation, predation by other biota, nutrient deficiency with consequent mortality, solar UV radiation, etc. In a study of *E. coli* pulse injected into a secondary effluent being discharged into a flowing stream, the microbes were observed after half an hour to decrease more than 10 times and by two hours to decrease more than 100 times (Palmeteer and Whitby, 1987). This would leave only 10^2 of the 10^4 functional mutants $.m^{-3}$. Similar reductions in other rivers during the summer were noted to take about 2 to 3 months and longer in the winter (Geldreich, 1989). The maximum survival of microbes in water or sediment is therefore limited. The $10^{-2}.m^{-3}$ functionally advantaged mutants are not therefore likely to accumulate.

(iii) The pathogens are shed by only 1 to 25% of the population and are not as abundant in effluents as the faecal coliforms. *Pseudomonas aeruginosa*, some strains of which can produce toxins, may approach the levels of the faecal coliforms, but others such as *Salmonella* spp. are typically less abundant than the faecal coliforms by 10 to 10^4 (Geldreich, 1989). The number of mutants of these pathogen species compared to mutants of faecal coliforms would also be expected to be fewer. This means less than 10 to 100 mutants $.m^{-3}$ and in some cases no mutants are likely to be created for some pathogens.

(iv) Pathogenicity does not generally result from contact with a single microbe, but some infectious dose is required to overcome the natural defence mechanisms of the human body. For some organisms such as *Shigella* spp. and enterovirus, this may be as few as 100 microbes; but for others such as *Salmonella* spp., it may take 10^6 or more microbes (Geldreich, 1989) for infection to result in clinical symptoms. The conservatively estimated production rate of 10-100 new mutants $.m^{-3}$ for a pathogen, even if consumed by a single individual, is 1 to 1×10^{-6} of the infectious dose depending on the pathogen involved.

The crude worst case considerations outlined above suggest that production of mutants by UV disinfection is essentially not a health risk. With fewer than 10-100 hypothetically advantaged mutant microbes of a given species being produced $.m^{-3}$, an infection dose of 100 microbes to 1 million being required for clinical symptoms, and the very slim probability that a recreational water user would acquire that infectious dose through breaks in the skin or by ingesting the necessary volume of water, it is not surprising that no mutant pathogen problems have been reported in UV disinfected waters. UV is used to disinfect drinking water in some large European cities in Holland, Germany, Austria and Switzerland and has been used in the latter two countries since 1955. By 1985 over 1000 installations were in use in these two countries alone (Kruithof and van der Leer, 1990). No mutant pathogen problems have been reported. Similarly, there have been no reported mutant pathogen outbreaks in hospital operating rooms bathed in UV treated air or in natural waters irradiated with solar UV for the past several centuries since society first began discharging waste into water ways.

Of equal importance, UV does not transform chemicals in water to generate mutagenic materials which could linger in the water after disinfection and have effects on receiving water biota or humans. Unlike chlorine, which can react with organics (in the water or within the organisms ingesting residuals) to generate chloroorganics having mutagenic effects (Kool *et al.*, 1985), UV does not produce elevated mutagen levels as judged by the Ames test even when treating heavily contaminated water such as Rhine river water exposed to 120 mW.s.cm^{-2} for 26 seconds (Zoeteman *et al.*, 1982). In contrast, chlorine revertants with different test strains using the same stored Rhine river water (Kool *et al.*, 1985). No elevated mutagenicity has been detected upon UV disinfection (6 -20 mW.s.cm^{-2}) in Holland of bank filtered Rhine water or stored surface Meuse river water (Kruithof and van der Leer, 1990). Chlorination did show an increase in mutagenicity in these waters.

Our discussion to this point has focussed on the changes in nucleic acid resulting from absorption of light below 300nm by the nucleic acids; however, UV light at longer wavelengths is found in solar irradiation and emissions from the medium and high pressure mercury arc lamps as well as other lamp sources. At these longer wavelengths a new mechanism of UV lethality, i.e. mutagenesis and carcinogenesis comes into play. This latter mechanism involves photosensitized (photodynamic) reactions in which reactive species of molecular oxygen are created when oxygen interacts with a light-excited naturally occurring photosensitizing molecule which has absorbed the UVA or UVB wavelengths (Peak *et al.*, 1984). The generally short-lived reactive oxygen species can bring about lethal, mutagenic, or carcinogenic chemical changes in nearby nucleic acids, but also can lethally transform other critical cellular constituents such as membrane molecules, enzymes, etc.

THE PROCESS OF UV DISINFECTION IN DELIVERING LETHAL UV DOSES TO THE MICROBES: PROCESS VARIABLES

The lethal effect of UV is a function of the dose of UV delivered to the microbe. Dose is defined as the product of the intensity of UV falling upon the microbe multiplied by the time interval over which that intensity is applied (i.e. dose = intensity x time). The number of photochemical transformations brought about in the nucleic acids is expected to be related primarily to the delivered dose of UV, but microbe mortality is a more complex function which incorporates (i) the dependence of the photochemical transformations on the nucleic acid structure, (ii) the number of transformations needed to inactivate the microbe, (iii) heterogeneity of response to UV within the irradiated microbial population and (iv) the operation of repair mechanisms, etc.. The typical response to a linear change in dose is a logarithmic change in the fraction of surviving microbes (Figure 12.6). In Table 12.1 we saw that different doses were required to achieve a given level of mortality with different microbes. This is equivalent to saying that the initial slope of the curve in Figure 12.6 varies between different microbes.

An important additional feature which characterizes the response of microbial populations (as opposed to individual cells) to UV and which is shown in the curve of Figure 12.6 is the departure from linearity at higher doses. This deviation results

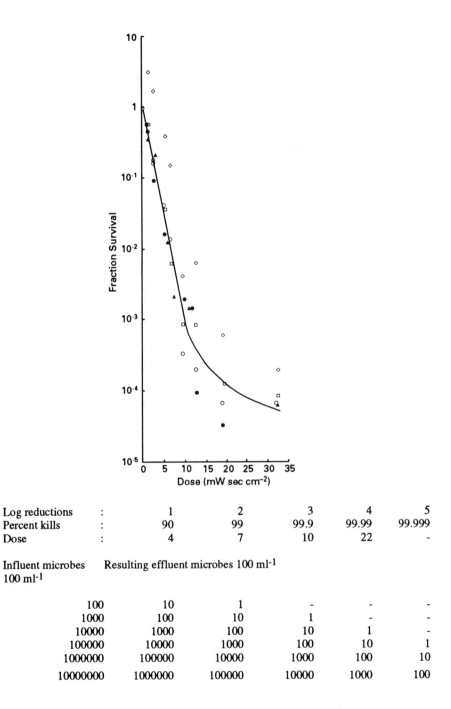

Log reductions	:	1	2	3	4	5
Percent kills	:	90	99	99.9	99.99	99.999
Dose	:	4	7	10	22	-

Influent microbes 100 ml^{-1}	Resulting effluent microbes 100 ml^{-1}					
100	10	1	-	-	-	
1000	100	10	1	-	-	
10000	1000	100	10	1	-	
100000	10000	1000	100	10	1	
1000000	100000	10000	1000	100	10	
10000000	1000000	100000	10000	1000	100	

Figure 12.6 – Fraction survival curve of UV irradiated faecal coliforms from wastewater.

from the natural or engineered tendency of microbes to aggregate and the consequential protection of microbes within the interior of the aggregates from receiving the same intensity of UV as that which falls upon the exterior of the aggregate. Since intensity within the aggregate is reduced, dose is also reduced and the average mortality within the aggregate decreases. The quality of the suspended solids entering the UV disinfection unit operation is very important in determining the lower limit of UV disinfection. The factors which determine the disinfectability of microbial populations containing particles (or more precisely, which determine the number of microbes which can not be inactivated within the core of particles) are:

(i)   the intensity of UV light falling upon the exterior of the particles,

(ii) the particle size distribution (larger particles will contribute the largest core volume, and hence number, of less readily reached microbes),

(iii) the total number of particles having the above particle size distribution, and

(iv) the optical density of the particles (particles in which the microbes are more densely packed due to the process of particle formation or the physiological state of the microbes at the time of aggregation, or particles which have incorporated UV absorbing organics or inorganics will not allow light to penetrate as deeply as less optically dense particles).

Several efforts have been made to quantify the impact of suspended solids on UV disinfection. Earlier studies, summarized in Stover et al., (1986), have attempted empirically to relate the limits of disinfectability to the total amount of suspended solids in the wastewater (Figure 12.7). Other studies (Qualls et al., 1985) have attempted to determine the limiting size for particle disinfection and within the limits of the methods used, found that the larger particle sizes (>20 or 40 microns) provide a better protection than smaller particles even though the latter are more abundant. No studies completed, to date, have incorporated particle size analysis as a parameter either to predict disinfectability of wastewater by UV or to assess the ability of different treatment processes (biological, sedimentation, coagulation/flocculation, filtration, etc.) to deliver desirable particle quality to the UV disinfection process. Trojan Technologies has, along with collaborating researchers, initiated such studies and will incorporate particle size distribution and particle optical quality studies into future projects assessing the application of its technology.

The intensity of light incident upon the surface of a particle or an individual microbe in the irradiated population is a function of:

(i)   the emission intensity of the light source used (this depends on the lamp chosen, its age and operating conditions),

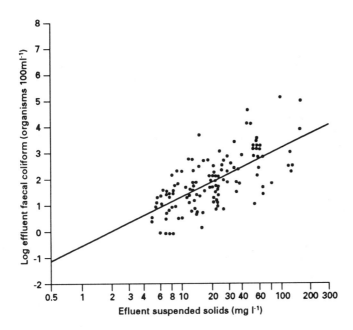

Figure 12.7 – Relationship between wastewater disinfectability and the level of suspended solids.

(ii) the reduction in intensity which might occur from a quartz protective sleeve surrounding the lamp and from any adherent coating of UV-absorbing molecules and ions built up on the sleeve,

(iii) the grid configuration and spacing of the lamps in the wastewater being treated,

(iv) the UV transmitting properties of the wastewater (this is usually measured by the percentage of transmittance of the wastewater at biocidally effective wavelengths such as 254nm),

(v) the thickness of the water layer through which the light must penetrate to reach the particle or microbe, and

(vi) the effectiveness of mixing which brings the particle or microbe into different intensity zones within the lamp grid during the particle's or microbe's course through the UV reactor (assuming perfect mixing, the particle or microbe actually receives an *average intensity* related to the time it spends within each intensity zone within the reactor).

The *average dose* delivered within the photoreactor is therefore a product of the *average intensity* and the *average retention time* within the reactor. Optimal reactor design must therefore consider not only UV intensity delivery, but also the hydraulic behaviour of the reactor. Optimally designed reactors provide both plug flow of fluid through the reactor to minimize differences in retention time for different volume elements, and good dispersion (turbulence) conditions to provide the lateral mixing necessary for (vi) above.

The UV equipment design engineer has a large measure of control over (i), (iii) (v) and (iv) above, but the wastewater characteristics are largely responsible for (ii) and (iv). Since the wastewater characteristics are constantly changing (with influent composition, with wastewater treatment process and the way the process is operated, with the hydraulic load through the treatment plant, etc.), and since changes continuously occur in the hydraulic load (which determines retention time within the reactor), it becomes important that the UV equipment engineer designs and tests his equipment for performance under worst case considerations (poorest wastewater quality expected to be encountered and the highest flow rate expected).

Evaluating the performance of UV disinfection equipment can be done through empirically evaluating separately the two components of dose: retention time distribution curves (Stover *et al.*, 1986) and the average intensity within the reactor; or more definitively, directly and preferentially through monitoring the reactor's ability to produce microbe mortality.

(i) If the objective of the performance evaluation is merely to determine that the equipment delivers the design dose under specified conditions, then a bioassay may be performed using non-aggregating microbes or spores whose dose-survival curve has been established in the laboratory (Stover *et al.*, 1986). The mortality achieved when the calibrated microbes or spores are passed through the photoreactor can then be used to read the actual dose delivered from the calibrated dose-survival curve and to compare this value with the dose which the equipment was designed to deliver under those specified conditions.

(ii) If the objective of the performance evaluation is to determine the applied dose (average intensity x retention time) and hence electrical power and cost necessary to achieve a target level of disinfection of a wastewater with previously undefined and changing characteristics (such as total suspended solids, particle size distribution, wastewater percent transmittance, particle optical density, etc.), then a pilot study which monitors wastewater microbe survival as a function of the changing wastewater characteristics (bulk properties and particle properties) and the applied dose parameters of intensity and retention time will provide the data for disinfection equipment design and scale-up.

UV disinfection should be considered an integral part of the overall process of wastewater treatment. The quality of the water entering the UV equipment and hence that equipment's performance is a function of all previous processing. The process design engineer who wishes to include UV disinfection as the last step in the process, or the operator who wishes to fine tune his process for UV disinfection, can optimize the pretreatment steps if they consider the impact of treatment on the final wastewater characteristics. The major factors influencing wastewater characteristics are, first, process variables which influence particle size distribution and the total amount of suspended solids in the final effluent including;

(i) nature of the bioprocess (simple aeration, activated sludge, biofilm process, etc.) as a function of:

- agitation and aeration practices which influence the ability of microbes to convert nutrients into self-aggregating microbial surfaces, and which influence the shear forces leading to particle disruption

- nutrient loading as monitored in various operational parameters such as F/M ratio, sludge age, influent BOD (COD) levels, etc. which can alter the nutritional and physiological state of the microbes and hence the aggregating and settling properties of the produced biomass. (Nutrient loads can vary with time of day, day of week, intermittently with industrial influent, or seasonally due to tourists, students or seasonal industrial influent)

- hydraulic loading which impacts on the process residence time, nutrient delivery, shear forces on biofilms and microbial aggregates, etc. (Hydraulic loads can vary with time of day, day of week, storm occurrence, etc.)

- industrial organics content of the influent and its impact on the bioprocess.

(ii) clarifier performance which is a function of

- clarifier design hydraulic load and/or the rate of change in hydraulic load

(iii) use of flocculants/coagulants to improve sedimentation and the

(iv) presence or absence of post-sedimentation filters;

In addition, a second group of process variables which influence the optical density of the bulk liquid and/or of the suspended solids in the bulk liquid. These include;

(i) presence of industrial content (e.g. dyes) which are soluble in the effluent or adsorbed onto the suspended solids,

ii) chemical additives for phosphate control (particularly iron) and the

iii) use of coagulants which remove UV absorbing colloids from the final effluent

A third group of process variables influence quartz sleeve coating and coating rate. However, the evolution of the UV industry is towards in-place cleaning mechanisms which will eliminate this problem. At the moment, the organics and inorganics which adhere to the sleeve must be removed periodically by physically wiping the sleeves individually (for small systems) or by immersion of the racks of lamps (modules), either individually or in multiples, into a bath containing an acid wash. The frequency with which this is required depends on the rate of coating buildup, which in turn depends on wastewater quality and flow rates. Cleaning monthly or biweekly is the most common experience for domestic wastewater.

UV DISINFECTION AS AN ALTERNATIVE TO CHEMICAL DISINFECTION: ADVANTAGES AND DISADVANTAGES.

Advantages:

- no production of toxic, carcinogenic or mutagenic chloroorganics such as the trihalomethanes (THM). The problems of chloroorganic production increase as the wastewater quality decreases. The chlorine consuming ability of the additional organics requires higher levels of chlorine for disinfection and consequentially results in formation of additional byproduct.

- no corrosive chemicals used in the plant and no toxic residuals being discharged to impact on the biology of the receiving water.

- no operator or neighbourhood safety risk due to transport to site, storage on site or use at site of toxic gases (chlorine for disinfection or sulphur dioxide for dechlorination) or caustic liquids (e.g. hypochlorite solutions). The safety risks are being translated into additional costs in the form of insurance and into additional ventilation and storage requirements for those communities which adopt the US Uniform Fire Code. The latter code requires separate storage buildings with their own exhaust, negative internal pressure, and facilities to handle the accidental release of gas by diluting, adsorbing, neutralizing, burning or otherwise processing the entire contents of the largest single tank or cylinder of gas stored. The code requires that the system shall be designed to handle the maximum anticipated pressure realized when equilibrium is reached. Such additional requirements could add an additional 30% to the cost of chlorination/dechlorination systems. The Uniform Fire Code is being enforced in Arizona, California, Minnesota and Washington in the USA. UV equipment does not even require an indoor installation.

- no long contact times required as with most chemical disinfectants. For treatment plants with limited space, UV disinfection can often be fitted whereas chemical contact chambers may not. The smaller space requirement for UV allows any available space to be used for expansion of other unit operations (biological, sedimentation, etc.). Unused chlorine contact chambers can be partitioned and partly used for UV and partly reused for other purposes.

- less dependence on microbe surface chemical structure and greater confidence in the relatedness of indicator organism mortality and pathogen mortality as discussed above.

- favourable capital and operating costs.

Disadvantages:

- UV disinfection equipment design and performance is based on wastewater quality parameters (e.g. percent transmittance at 254nm, particle size distribution, etc.) not typically monitored in the wastewater industry and for which only a limited data base exists. Trojan Technologies is attempting to correct that situation and to help design engineers realize how to interface design and operation of UV disinfection with other unit operations. Particle size distribution has, for example, application in design and performance of coagulation/flocculation processes, sedimentation and filtration, all of which can be used as pretreatments prior to disinfection.

UV DISINFECTION - THE CURRENT TECHNOLOGY

In a survey done by the USEPA in 1986, approximately 50 UV systems had been installed in the wastewater treatment plants in North America (White et al., 1986). Trojan Technologies alone now has over 500 such installations, most of which are at treatment plants which had previously used gas chlorination. In the majority of cases, the UV systems were retrofitted into the existing chlorine contact chamber. The technology has now been introduced into Europe, Australia and Asia.

Most UV disinfection systems used for treatment of secondary effluent quality wastewater (less than 30 mgl^{-1} suspended solids) employ low pressure mercury arc lamps which produce more than 80% of their emission at 254nm. The lamps resemble fluorescent lamps except they lack the fluor and have a UV transmitting quartz envelope. The lamps are powered by a conventional or electronic ballast (a circuit component which controls the current passing through the lamp). A quartz sleeve surrounds the lamp to protect the lamp and to provide a working temperature closer to the optimum for lamp UV output.

Figure 12.8 illustrates how UV can be retrofitted into an existing chlorine contact tank, and the approximate savings in space. This particular installation was in a building, although the UV equipment did not require it. Figure 12.9 shows the modular concept which has been used in UV equipment design for wastewater treatment since Trojan Technologies introduced the open channel modular system onto the market in the early 1980s. Each rack of lamps (one module) can contain 2 to 16 lamps, and each module is connected by a cable to its own power and control module in the control panel. Several lamp modules side by side in a channel make up a bank of modules. Flow is parallel to the lamps. An automatic level controller at the end of the channel after the UV modules keeps the water at the desired level above the lamps. The modular concept allows treatment plants to match the disinfection power requirements to changing demand resulting from fluctuating flow rates through the plant. Since low pressure mercury arc lamps can not be modulated in output as can medium pressure mercury lamps, power modulation is accomplished by using several parallel channels with a bank of lamps (or more commonly two banks in series within each channel). As the flow rate to the plant increases, additional channels and banks of lamps are brought on stream. The modularity also allows planning for the future by allowing engineers to build channels with temporary void filling structures which can be removed and replaced with lamp modules as the need for expansion is realized.

Several case studies of UV installations at municipal secondary wastewater treatment plants have recently been published (Maarschalkerweerd et al., 1990). Performance, capital and operating costs were reviewed. The data in Figure 12.7 indicate UV performance as a function of effluent suspended solids. Data currently being collected suggests that some of the scatter can be related to impact of the wastewater treatment process on particle quality.

The consulting design engineers for the Northwest Bergen County Utilities Authority have recently published (Fahey, 1990) a comparison of projected operating and maintenance costs using different technologies to disinfect the secondary

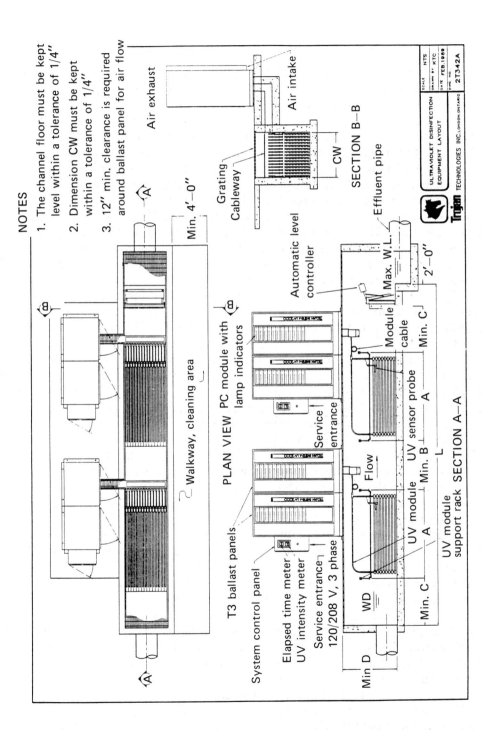

Figure 12.8 – Retrofit of UV into an existing chlorine contact tank at Northwest Bergen County's wastewater plant.

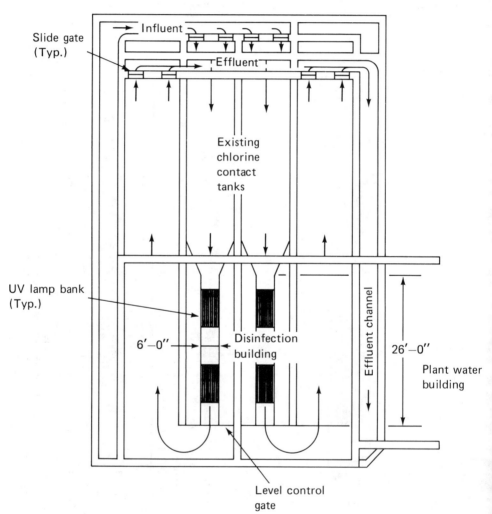

Figure 12.9 – Equipment layout showing modules (racks of lamps) placed in parallel and series.

effluent of the Waldwick, New Jersey, USA plant (24 Million US Gallons per Day average flow or 90800 m^{3}d^{-1}). The alternatives and annual costs (US$) were as follows:

    (i) liquid gas chlorination, sulphur dioxide dechlorination   107,000

    (ii) liquid gas chlorination, sodium bisulphite dechlorination 110,000

    (iii) hypochlorination with sulphur dioxide dechlorination   115,000

    (iv) hypochlorination with sodium bisulphite dechlorination   119,000

    (v) ultraviolet   65,000

    (vi) ozone   162,000

The design engineers selected UV on the basis of cost and its ability to meet the more stringent permit levels for chlorine. Since conversion from chlorine to UV in 1989, the plant has been achieving 20 times better than the effluent permit which is 200 faecal coliforms .100ml^{-1}. The plant spends only 60% as much personnel time on maintaining the UV system as it did with chlorine. The overall annual operating cost is $2700 US/MGD ($716/1000 m^{3}d^{-1}) for this sized installation where power costs $0.06 kwh.

Capital costs for open channel low pressure mercury arc lamp UV systems can vary somewhat from site to site depending on installation size, water quality, shipping costs, tariffs, etc.; however Table 12.3 provides some guidance for capital costs in North America (US$) for installations of different sizes:

Table 12.3 – Capital cost for plant of different sizes.

Plant size in MGD (m^{3}d^{-1})	$US/MGD
less than 1 MGD (less than 3785 m^3/d)	50,000
1 - 3 (3785 - 11355 m^{3}d^{-1})	40,000
3 - 7 (11355 - 26495 m^{3}d^{-1})	35,000
7 - 10 (26496 - 37850 m^{3}d^{-1})	30,00
> 10 (> 37850 m^{3}d^{-1})	25,000

An estimate of the costs for a 26.4 MGD peak flow (1000,000 m^{3}d^{-1}) system in the United Kingdom was provided by Mr. Peter Daniels of Sunwater Limited, Droitwich. The below prices are for a good quality secondary effluent (20 ppm suspended solids) and a targetted effluent standard of 200 faecal coliforms 100 ml^{-1}, but do not include engineering costs for site preparation:

Capital cost (Pounds)	£350,000
Average annual lamp cost (Pounds)	£12,000
Average annual power cost (Pounds)	£25,000
Operating costs (lamps + power)	0.3 pence/m^3
Capital costs (over 20 years)	0.3 pence/m^3

In summary, UV disinfection has proven to be an environmentally and community safe, operator friendly, effective and economically viable method of protecting receiving waters from pathogens being discharged from wastewater treatment plants. Such protection is in the best interests of the public which may use the waters for recreational purposes and/or draw its drinking water from such sources.

REFERENCES

Bailey, C.A., Neihof, R.A. and Tabor, P.S. (1983) Inhibitory effect of solar radiation on amino acid uptake in Chesapeake Bay bacteria. *Applied and Environmental Microbiology 46*, 44-49.

Brash, D.E. (1988) UV mutagenic photoproducts in *Escherichia coli* and human cells: a molecular genetic perspective on human skin cancer. *Photochemistry and Photobiology 48*, 59-66.

Calkins, J., Selby, C. and Enoch, H.G. (1987) Comparison of UV action in spectra for lethality and mutation in *Salmonella typhimurium* using a broad band source and monochromatic radiations. *Photochemistry and Photobiology 45*, 631-636.

Chang, J.C.H. *et al.* (1985) UV inactivation of pathogenic and indicator microorganisms. *Applied and Environmental Microbiology 49*, 1361-1365.

Fahey, R.J. (1990) The UV effect on wastewater. *Water Engineering and Management 137 (12)*.

Geldreich, E.E. (1989) Microbial water quality concerns for water supply use. *Microbial Water Quality Concerns for Water Supply Use* Water Pollution Control Federation Specialty Conference *Microbial aspects of surface water quality*, Chicago, Illinois.

Harris, G.D. *et al.* (1987) Ultraviolet inactivation of selected bacteria and viruses with photoreactivation of the bacteria. *Water Research 21*, 687-692.

Hutchinson, F. (1987) A review of some topics concerning mutagenesis by ultraviolet light. *Photochemistry and Photobiology 45*, 897-903.

Kool, H.J. *et al.* (1985) *Mutagenic and Carcinogenic Properties of Drinking Water* In Jolley, R.L. *et al.* (eds) *Water Chlorination. Chemistry, Environmental Impact and Health Effects* 5. Lewis Publishers, Inc. , Chelsea, Michigan.

Kruithof, J.C. and van der Leer, R.C. (1990) *Practical Experiences with UV-Disinfection in the Netherlands.* Proceedings of the AWWA Seminar on Emerging Technologies in Practice. Cincinnati, Ohio. American Water Works Association, Denver Colorado.

Maarschalkerweerd, J., Murphy, R. and Sakamoto, G. (1990) Ultraviolet disinfection in municipal wastewater treatment plants. *Water Science Technology 22*, 145-148.

Meckes, M.C. (1982) Effect of UV light disinfection on antibiotic resistant coliforms in wastewater effluents. *Applied and Environmental Microbiology 43*, 371-377.

Murray, G.E. *et al.* (1984) Effect of chlorination on antibiotic resistance profiles of sewage-related bacteria. *Applied and Environmental Microbiology 48*, 73-77.

Ossanna, N., Peterson, R.R. and Mount, D.W. (1987) UV inducible response in *Escherichia coli*. *Photochemistry and Photobiology 45*, 905-908.

Palmateer, G. and Whitby, G.E. (1987) *Ultraviolet Disinfection: Its effect on Escherichia coli and Bacteriophages as indicators of disinfection efficiency of wastewater.* Technology Transfer Conference, Toronto, Ontario.

Peak, M.J. *et al.* (1984) Ultraviolet action spectra for DNA dimer induction, lethality and mutagenesis in Escherichia coli with emphasis on the UVB region. *Photochemistry and Photobiology 40*, 613-620.

Pettibone, G.W., Sullivan, S.A. and Shiaris, M.P. (1987) Comparative survival of antibiotic-resistant and sensitive fecal indicator bacteria in estuarine water *Applied and Environmental Microbiology 53*, 1241-1245.

Qualls, R.G. (1985) Factors controlling sensitivity in ultraviolet disinfection of secondary effluents. *Journal of the Water Pollution Control Federation 57*, 1006-1011.

Scheible, O.K., and Bassel, C.D. (1981) *Ultraviolet disinfection of a secondary wastewater treatment plant effluent.* EPA - 600/2-81- 152. USEPA, Cincinnati, Ohio, USA.

Smith, A.W., Davies, D.J.G. and Moss, S.H. (1987) Photoreactivation of ultraviolet radiation damage in dark repair deficient PHR mutants of *Escherichia coli* K-12. *Photochemistry and Photobiology 45*, 247-252.

Stover, E.L. *et al.* (1986) *Design Manual. Municipal Wastewater Disinfection.* EPA-625/1-86/021. USEPA, Cincinnati, Ohio, USA.

Whitby, G.E. *et al.* (1984) *Journal of the Water Pollution Control Federation 56*, 844-850.

White, S.C., Jernigan, E.B. and Venosa, A.D. (1985) *A study of operational ultraviolet disinfection equipment at secondary treatment plants.* Presented at the 58th Annual Conference, Water Pollution Control Federation, Kansas City, Mexico.

Yip, R.W. and Konasewick, E. (1972) Ultraviolet sterilization of water - its potential and limitations. *Water and Pollution Control (Can) 110(6)*, 14-18.

Zoeteman, B.C.J. *et al.* (1982) *Environmental Health Perspect. 46,* 197-205.

Chapter 13

SUMMARY AND CONCLUSIONS

Frank Jones and David Kay
Centre for Research into Environment and Health, University of Wales, Lampeter SA48 7ED

THE PROBLEM

The central problem in many environmental issues is that they trespass traditional scientific and academic disciplinary boundaries. Recreational water quality management is a classic example of just such a problem. Too many attempts to define this issue and present solutions have depended on single disciplinary approaches which have run into significant problems because the researchers have not taken sufficient account of the breadth of expertise required to produce robust scientific information on which to base policy. The science involved stretches current boundaries of each of the disciplines involved which stimulates and interests the researcher but can result in a lack of clarity and some disillusion at the level of the policy maker.

THE SCIENTIFIC DEBATE

The central scientific question remains, whether bathing and other recreational activities in waters contaminated by sewage results in measurable health effects in the exposed population. If the answer to this question is affirmative, then the policy maker requires information on the relationship between pollution levels and associated morbidity in the recreator population. These data, in the form of a dose-response relationship, can then be used to predict the health impacts of differing pollution levels and, thus, define appropriate standards for recreational waters.

Expressed in these terms, the problem seems simple. However, previous attempts to resolve this scientific question have been plagued by the oversimplification of the problem and produced results which do not stand up to scrutiny. Fleisher's analysis presented in Chapter 9 illustrates this problem. He has cast doubt on the science which forms the foundation for current US Federal water quality criteria for recreational waters. This harsh evaluation of the research conducted by Cabelli *et al.* (1982) is very similar to that presented by Cabelli *et al.* (1975) when criticising earlier work of Stevenson (1953) which underpined the US standards then in force. In the United States, therefore, we can see two clear phases of research in which first Stevenson (1953) and then Cabelli *et al.* (1982) addressed the problem with the aim of providing relevant policy information. The rationale for both studies was that the existing science was inadequate and each new phase of activity was characterised by significant methodological development in the prospective epidemiological methods employed. Fleisher's critique may indicate the initial stages of a third phase of research activity. This has significance well beyond North America because many other nations have adopted the epidemiological techniques pioneered by Cabelli *et al.* (1982) and similar problems may be evident in other studies (Cheung *et al.*, 1988; Fattal *et al.*, 1986; El Sharkawi *et al.*, 1982; New Jersey Health Department, 1988,1989; Seyfried *et al.*, 1985a,b). There is certainly some justification for the re-examination of these studies in the light of Fleisher's comments and those of Lightfoot (1989)

In Europe, a more confused pattern of research methods and timing is evident. This reflects the heterogeneous nature of recreational water quality management in the European context. Marine epidemiological research dates from the PHLS (1959) investigations which were retrospective studies of notifiable (i.e. serious) disease. Cross sectional studies have been completed in Spain (Mujeriego *et al.*, 1983) and the United Kingdom (Brown *et al.*, 1987). Prospective studies using diary sheets were used in France (Foulon *et al.*, 1983). This lack of 'European' research coordination is understandable given the 'national' nature of these investigations but it severely limits the utility of the resultant information. Coordination and integration of the European data is essential and must be achieved if the present or revised standards, defined in the EC Bathing Water Directive (76/160/EEC), are to have credibility in all EC member countries (EEC, 1976).

The record of the United Kingdom in recreational water quality research has been characterised by periods of intense activity interspersed with quiescent apathy. The initial PHLS (1959) investigations have been much criticised in the subsequent period but, in fact, they represented a carefully conducted piece of research directed at a very narrow question. The question was whether poliomyelitis and enteric fever could be contracted by bathing in sewage contaminated sea water. The finding that, for all practical purposes, this risk could be ignored represented a credible and scientifically consistent outcome for this investigation. It could be argued that the PHLS chose to address the wrong question. Particularly since Stevenson (1953) had suggested that less serious ailments might be contracted by bathing in sewage contaminated waters. Nearly thirty years elapsed before this question was addressed by the United Kingdom research community in a concerted national research programme (Chapters 7 and 10 this volume). The initiation of this current research programme in the United Kingdom presents parallels with the critique of previous work and significant methodological advances achieved in the United States by Victor Cabelli. The critique came in the mid 1980s when public awareness of this issue was heightened by considerable pressure group interest and Parliamentary attention (Jones et al., 1990; Kay and McDonald; 1986; HMSO, 1985a,b,c). One pressure group (Greenpeace UK Ltd) sponsored research which maintained this pressure and public interest (Brown et al., 1987). Methodological advances, in terms of significant enhancement of the methods proposed by Victor Cabelli and the use of the volunteer cohort approach, have now been achieved in this research programme (Jones et al., 1991; Pike, 1991, Chapters 7 and 10 this Volume). The recent parliamentary enquiry Chaired by Sir Hugh Rossi represents a significant milestone in this debate and its recommendation (Chapter 1 this volume) reinforced the commitment to significant enhancements of recreational water quality in United Kingdom sea waters (HMSO, 1990).

THE POLICY RESPONSE

In the United Kingdom, the research outlined above should produce credible scientific information on which to base policy for the management of marine recreational waters by the early part of 1993. The National Rivers Authority is responsible for suggesting appropriate policies to the Secretaries of State for the Environment and its recent consultative and policy documents (NRA, 1991a,b) outline a sensible procedure for translating the scientific research into credible standards (Chapter 3 this volume). Two principles outlined in NRA (1991b) are central to credible standards formulation and are worthy of support by recreational interest groups and the scientific community. The first it is that the standards set for water contact activities should be so designed to protect those engaged in such activities (NRA, 1991b:19). This implies that the standards will be based on health-related criteria, presumably those derived from the national UK sea bathing research programme (Chapters 7 and 10 this volume). The second principle relates to a hierarchical classification of water contact activities which implies that a family of standards might be required to protect the spectrum of possible users. Both elements, which are currently in a consultation phase are scientifically consistent with the results of current research. The NRA intend to provide advice to the Secretaries of State on appropriate standards for marine recreational waters and wish to see this advice implemented by the winter of 1994 (NRA, 1991b:48). This timescale should be achievable in scientific terms, given the completion of present research programmes. However several policy questions will require resolution in the medium term. These include;

(i) the potential conflict between the Statutory Water Quality Objectives and any EC standards then in force (whether the present or some revised EC standards). Such a discrepancy is likely if the NRA decides to define a new set of criteria based on the current health research which could be used to protect those engaged in water contact activities. This objective could not be guaranteed with present EC standards which do not have a firm basis in epidemiological research. It would seem sensible to utilise the most stringent standard if such a discrepancy became evident.

Significant expenditures, of over £2.5 billion, are currently being devoted to achieving the present EC Imperative standards at UK identified bathing waters. Given the stated intention of the NRA to develop a parallel, and potentially different, set of standards by the spring of 1994, it would be wise for present disposal schemes to maintain a flexible approach at the initial design stage and build in a considerable margin for potential improvement if so required by future standards. The dilemma facing the constructors of current schemes is that a design philosophy of continued flexibility with considerable safety margins would have significant cost implications which, at present (February 1992), could not be justified on other than speculative grounds.

(ii) It is likely that there will always be some measurable health effect attributable to bathing in sea water even with barely detectable pollution levels. This 'exposure' effect may be due to the fact that sea water is not the natural environment for the human body (Chapter 7 this volume). This exposure effect is difficult to disentangle from the direct health effects of pollution. It is likely, therefore, that whatever standards are defined for marine recreational waters, some risk of minor ailments will still be attributable to the act of bathing or other water activities.

The Statutory Water Quality Objectives set by the Secretaries of State are therefore unlikely to guarantee zero health effects. The residual morbidity rate for gastrointestinal symptoms has been termed 'acceptable' in the 1986 USEPA standards. Acceptable GI is an interesting concept but one which is unlikely to find favour amongst the general public who wish to know whether their bathing waters are, or are not, 'safe'. Given this scientists' perception of a spectrum of risks and the public perception of absolutes in which beaches either PASS or FAIL, a degree of public education will be required during the implementation phase of future standard systems.

(iii) As principal monitoring and control agency, the NRA is in a difficult position. It has no specific responsibility for health related matters but it is committed to the development of standards requiring epidemiological information for the protection of public health. Local responsibility for public health maintenance rests with District Council environmental health departments who are often faced with enquiries from potential recreators wishing to know if they might suffer significant health effects caused by bathing in marine waters. The front line responsibilities of environmental health officers is further reinforced by their role within local government which responds quickly to any suggestion of pollution at a resort town beach affecting the tourist trade with significant impacts on municipal economies. Close liaison between these two professional groups will be required during the implementation phase of any new standard systems because the local environmental health professionals will be faced with the main burden of public liaison and education.

(iv) The technology of waste water treatment is advancing rapidly and relatively 'clean' options now exist for the reduction in microbial populations and other pollution in effluents discharged to coastal waters (Chapters 6 and 7 this volume). It has been common practice in the UK for the opponents of coastal sewage disposal schemes to call for 'full sewage treatment' in the apparent belief that this panacea will solve the problem. Such calls often derive from environmental concern and the gut feeling that we should not be dumping our waste in the sea. However, there will be an environmental cost to all disposal options. For example, land-based tertiary treatment and disinfection will reduce marine microbial and BOD loadings but can also produce problems of sludge disposal and potential toxic input to the marine environment. We have very little environmental information on these matters at this time and urgent study is required on the associated environmental impacts of current treatment options.

(v) A very significant policy question raised by the NRA (1991b) is the development of standards or objectives for fresh recreational waters. Whilst not the central focus of this volume, it does raise a host of environmental, public health and legal matters (Chapter 5 this volume) which will be addressed over the next few years. Again significant divergence between existing policy in the United Kingdom, which has not identified fresh recreational waters for monitoring under the terms of the Bathing Water Directive (76/160/EEC), and the objectives of the NRA in setting water quality objectives for recreational activities in fresh waters is now evident. Hopefully, this will be addressed by designation or identification of appropriate fresh water sites under the terms of Directive (76/160/EEC) in due course.

The whole problem of fresh recreational water management and control is contentious and under-researched in Europe (Fewtrell, 1991). The legal and administrative problems presented by fresh water sites are complex and differ in many respects from those of marine locations (Chapters 5 and 6 this volume). Volume II in this series will address this problem in more detail.

As we approach the last decade of this century, the United Kingdom is beginning to address the problems of coastal sewage pollution which have grown during the previous 200 years. This book has shown the multidisciplinary nature of these problems. Their resolution requires inputs from politicians, legislators, environmental microbiologists, clinical microbiologists, medics, epidemiologists, statisticians, environmental health professionals, treatment technologists and marine scientists. It is, perhaps, the successful combination of these skills into a coordinated and cohesive research effort that has produced such rapid progress on this problem since the mid-1980s. Much remains to be done, however, both in terms of basic research and the successful translation of these research studies into practical environmental policy.

REFERENCES

Brown, J.M., Campbell, E.A., Rickards, A.D. and Wheeler, D. (1987) The public health implications of sewage pollution of bathing water. SWIRLS The University of Surrey Water Investigation and Research Service. Robens Institute, University of Surrey. Nov.

Cabelli, V.J., Dufour, A.P., McCabe, L.J. and Levin, M.A. (1982) Swimming associated gastroenteritis and water quality. *American Journal of Epidemiology 115(4)*, 606-616.

Cabelli, V.J., Levin, M.A., Dufour, A.P. and McCabe, L.J. (1975) *The development of criteria for recreational waters*. In Gameson, A.L.H. (Ed.) *Discharge of sewage from sea outfalls*. Pergamon. 63-74.

Cheung, W.H.S., Kleevens, J.W.L., Chang, K.C.K. and Hung, R.P.S. (1988) Health effects of beach water pollution in Hong Kong. *Proceedings of the Institution of Water and Environmental Management*. Proceedings of the Annual Conference 376-383.

EEC (1976) Council Directive of 8 December 1975 concerning the quality of bathing water (76/160/EEC). *Official Journal L/31*, 1-7.

El Sharkawi, F. and Hassan, M.N.E.R. (1982) The relation between the state of pollution in Alexandria swimming beaches an the occurrence of typhoid among bathers. *Bull. High Inst. Pub. Hlth. Alexandria 12*, 337-351.

Fattal, B., Peleg-Olevsky, T. Agurshy and Shuval, H.I. (1986) The association between sea-water pollution as measured by bacterial indicators and morbidity of bathers at Mediterranean beaches in Israel. *Chemosphere 16(2-3)*, 565-570.

Fewtrell, L. (1991) Freshwater recreation a cause for concern. *Applied Geography 11(3)*, 215-226.

Fleisher, J.M. (1991) A reanalysis of data supporting US Federal bacteriological water quality criteria governing marine recreational waters. *Research Journal of the Water Pollution Control Federation 63(3)*, 259-265.

Foulon, G., Maurin, J., Quoi, N.N. and Martin-Bouyer, G. (1983) Etude de la morbidite humaine en relation avec la pollution bacteriologique des eaux de baignade en mer. *Revue Francaise des Sciences de L'eau 2(2)*, 127-143.

H.M.S.O. (1984) *Royal Commission on Environmental Pollution Tenth Report. Tackling Pollution-Experience and Prospects.* Cmnd. 9149 H.M.S.O. London. 233p.

H.M.S.O. (1985a) *House of Commons Committee on Welsh Affairs Coastal sewage pollution in Wales.* Minutes of Evidence 16.1.85. H.M.S.O. London. 118-167p.

H.M.S.O. (1985b) *House of Commons Committee on Welsh Affairs Coastal sewage pollution in Wales.* Report and Proceedings Vol I and II. 12.12.85. H.M.S.O. London. 27p.

H.M.S.O. (1985c) *House of Commons Committee on Welsh Affairs Coastal sewage pollution in Wales* Minutes of Evidence 5.12.84. H.M.S.O. London. 117p.

H.M.S.O. (1990) *House of Commons Environment Committee. Fourth Report.* Pollution of Beaches Volume I. H.M.S.O. London 11 July 1990. lvii pp

Jones, F., Kay, D., Stanwell-Smith, R. and Wyer, M.D. (1990) An appraisal of the potential public health impacts of sewage disposal to UK coastal waters. *Journal of the Institution of Water and Environmental Management 4(3)*, 295-303.

Jones, F., Kay,D., Stanwell-Smith, R. and Wyer, M.D. (1991) Results of the first pilot scale controlled cohort epidemiological investigation into the possible health effects of bathing in seawater at Langland Bay, Swansea. *Journal of the Institution of Water and Environmental Management. 5(1)*, 91-98.

Kay, D and McDonald, A.T. (1986) Coastal Bathing Water Quality. *Journal of Shoreline Management 2*, 259-283.

Lightfoot, N.E. (1989). *A prospective study of swimming related illness at six freshwater beaches in Southern Ontario.* Unpublished Ph.D Thesis. 275p.

Mujeriego, R., Bravo, J.M. and Feliu, M.T. (1982) Recreation in coastal waters public health implications. *Vier Journee Etud. Pollutions, Cannes, CIESM.* 585-594pp.

New Jersey Department of Health (1988) *A study of the relationship between illness in swimmers and ocean beach water quality.* New Jersey Health Dept. March 1988. 167p.

New Jersey Department of Health (1989) *A study of the relationship between illness in swimmers and ocean beach water quality interim summary report.* New Jersey Health Dept. March 1989. 54p.

NRA (1991a) NATIONAL RIVERS AUTHORITY *Bathing Water Quality in England and Wales.* Report of the National Rivers Authority. Water Quality Series No 3. National Rivers Authority, Bristol. 188p.

NRA (1991b) NATIONAL RIVERS AUTHORITY *Proposals for Statutory Water Quality Objectives.* Report of the National Rivers Authority. Water Quality Series No 5. National Rivers Authority, Bristol. 188p

P.H.L.S. (1959) PUBLIC HEALTH LABORATORY SERVICE. Sewage contamination of coastal bathing waters in England and Wales: a bacteriological and epidemiological study. *Journal of Hygiene, Cambs. 57(4)*, 435-472.

Pike, E.B. (1991) *Health effects of sea bathing (EM 9511) Phase II studies at Ramsgate and Moreton. 1990.* Water Research Centre Report No DoE 2736-M. Water Research Centre, Medmenham.

Seyfried, P.L., Tobin, R., Brown, N.E. and Ness, P.F. (1985a) A prospective study of swimming related illness. I Swimming associated health risk. *American Journal of Public Health 75(9)*, 1068-1070.

Seyfried, P.L., Tobin, R., Brown, N.E. and Ness, P.F. (1985b) A prospective study of swimming related illness. II Morbidity and the microbiological quality of water. *American Journal of Public Health 75(9)*, 1071-1075.

Stevenson, A.H. (1953) Studies of bathing water quality and health. *American Journal of Public Health 43*, 529-538.

USEPA. (1986) UNITED STATES ENVIRONMENTAL PROTECTION AGENCY. *Ambient water quality criteria for bacteria - 1986*. EPA440/5-84-002. Office of Water Regulations and Standards Division. Washington DC 20460. 18pp.

Alphabetical Subject Index

A

A 'large number' of bathers   53
Abdominal cramp   74
Aberystwyth   5, 49
Acceptable swimming associated
    gastroenteritis   124, 137
Activated sludge   109, 191
Activated sludge effluent (ASE)   161
Adenovirus   78, 94
Adequate precision   111
Adhesion of bacteria   109
Aerobic genera   91
Aerobic staphylococci   91
Aeromonas hydrophila   74
Aeromonas spp.   72
Aetiology of minor symptoms   98
Aggregates   188
Aggregation of bacteria   109
Agitation and aeration   191
Agricultural waste   71
American Joint Committee on Bathing
    Places   90
Ames test   184
Amoebae   76
Analogy   95
Analysis of variance   105, 119
Analytical quality control   105
Anecdotal evidence of paratyphoid
    infection   131
ANOVA   7, 110
Antibiotic resistance   184
Apocrine glands   91
Arizona   193
ASE. *See* Activated sludge effluent (ASE)
Aspergillus niger   181
Aspirational pneumonia   74
Assessing compliance   106
Australia   159, 160, 164
Austria   185
Autochthonous   74
Average dose   190
Average retention time   190

B

Bacillery dysentery   73
Bacillus anthracis spores   181
Bacillus subtilus spores   181
Back River Wastewater Treatment
    Plant   161
Bacteraemia   73
Bacteria   181
Bacterial enteritis   74
Bacterial virus   168
Bacteriological water quality criteria   121
Bacteriophage   148, 149, 181
Baltimore Bureau of Wastewater   172
Baltimore, Maryland   161
Barely acceptable beaches   135
Bather cohort   142
Bather/non-bather differentials   143
Bathing activity   135
Bathing water   23, 52
Bathing Water Directive   19, 42, 52,
    57, 80, 90, 98
Bathing water quality   130
Beach interview   135
Beach survey   95
ß haemolytic streptococci   91
Bile salt   108
Bilthoven, The Netherlands   111
Bioassay   190
Biochemical oxygen demand
    (BOD)   159, 160
Biofilms   191
Biological gradient   95
Blackheath CMF   159
Blackpool   54
Blue Flag   35, 38, 71
Blue green algae   79
Blue Mountains of Australia   159
BOD. *See* Biochemical oxygen demand
    (BOD)
Boston, Massachusetts   7, 113,
    117, 120
Bottom swimming   76

Brackish waters 113
Brighton 54
Bristol 75
Britain 97
British Tourist Authority 34
Broad Street pump 89
Brooklyn 6
Bulk properties 190
Byelaws 53

C

Cabelli 15, 36, 90, 95, 113, 117, 118, 135, 148, 202
Cairns 6
California 193
Campylobacter spp. 72, 74, 143, 147
Canada 6, 96, 134, 140, 175
Candida albicans 72
Carbohydrate 178
Cartwright 6
Casitone-glycerol-yeast extract agar 109
Causal relationships 90
Causality between environmental exposure and disease 7
Central tendency 106
Centre for Law in Rural Areas 5, 49
Centre for Research into Environment and Health 5, 6, 71, 129
Chain of infection 92
Chartered Institute of Public Finance and Accountancy 37
Chemical disinfection 80
Chi^2 145
Chlorinated hydrocarbons 148
Chlorination 184, 186
Chlorine 74, 178
Cholera 73, 77, 89, 147
Chronic illness 143
Clarifier performance 191
Clinical analysis of samples 146
Clostridia spp. 72
Clostridium perfringens 169
Clostridium tetani 181
CMF. See Continuous microfiltration (CMF)

Coastal waters 59
Codes of practice 58
Coherence 95
Coliform bacteria 23, 146
College of Commissioners 22
Colour 20, 27, 29
Colour-blind scientists 110
Commission of the European Communities v. Kingdom of the Netherlands 51
Committee on Adaptation to Technical progress 28
Commons Environment Committee 36
Commons Welsh Affairs Committee 34
Communicable Disease Centre (Scotland) 94
Communicable Disease Surveillance Centre 94
Community Law 50
Compliance 26, 28
Conflict of philosophies 99
Conjunction 52, 65
Conjunctivitis 78
Consents 63
Consistency 95
Contact person 139
Contaminated drinking waters 96
Continuous microfiltration (CMF) 157, 159
Controlled waters 59
Corynebacterium diptheriae 181
Council of Ministers 52
Counting of colonies 108
Court of Justice. See European Court of Justice
Coxsackie viruses 78, 148, 178, 180
Crab meat 77
Crown Estate Commissioners 56
Crunulla in New South Wales 164
Cryptosporidia 73, 74, 147, 149
Cryptosporidiosis 74, 94
Cyanobacteria 79
Cyanobacterial poisoning 79
Cyclobutadithymine dimer 178
Cyclobutane dimers 178

Cyst   147
Cytosine   178

D

Dangerous Substances Directive   55, 61
Decay rate   182
Delaying tactics   71
Density of bathers   55
Department of the Environment   97
Department of the Interior   132
Dependent variable   115
Depot or reservoir   93
Derogations   54
Designate   52
Designated bathing waters   54
Designation   65
Devon-Dorset coast   77
Diarrhoea   74, 77, 98
Dietary habits   143
Dilution of effluent   184
Dilution of samples   107
Diptheroids   91
Directorate General XI   5
Dirty beaches   71
Discharge consents   63
Disinfection   23, 29, 74, 157, 190
DNA   176, 180
DNA photoproducts   179
Dogs   79
Dose-response relationships   135, 136
Dose-survival curve   190
Draft Directive on Municipal Waste Water. *See* Municipal Waste Water Directive
Drinking water   198
Drowning   74
Dual-temperature, time-cycling incubators   108

E

Ear and skin infections   148
Ear and throat swabs   143
Ear infections   77
Early standards   132
EC Bathing Water Directive   15, 16, 26, 40

Eccrine sweat glands   91
Echovirus   78
Economic and Social Committee   21
EEC   19
EEC Treaty   21
Effluent   62
Effluent channel   182
Egypt   96, 140
Encephalitis   78
Endotoxic shock   77
Endotracheal aspirates   74
England   56, 94, 130
English Tourist Board   34
Enteric fever   73
Enteric viruses   28
Enterococci   114
Enterovirus   78, 80, 90, 98, 147, 148, 161, 185
Environment Committee   16
Environmental Assessment Directive   55
Environmental Health   5
Environmental Health Officers   35, 142
Environmental policy   206
Environmental pressure groups   35
Environmental Protection   33
Environmental Quality Objectives   40
Environmental Quality Standards   40
Environmentally resistant microbial indicators   136
Enzymatic methods   80
Epidemic vomiting and diarrhoea   78
Epidemiological evidence   80
Epidemiological research   6, 28, 29, 129
Epidemiological risks   135, 142
Epidemiological studies   44
Escherichia coli   73, 79, 178, 181, 184
Eskimo roll   37
Eurobeaches   34
European Commission   16, 65
European Communities Act (1972)   50
European Court of Justice   19, 50
European Parliament   21
European Standards   134
Eutrophication   148

Exceptional weather 55
Exclusive Economic Zone 56
Experiment 95
Exposure effect 204
Eye symptoms 98
Eyes 92

F

F^2 bacteriophage 81
Faecal coliform/faecal streptococci ratio 79
Faecal coliforms 42, 90, 160, 164
Faecal sample 143, 147
Faecal streptococci 79, 80, 90, 96, 98, 108, 161
Faecal-oral route 77, 96
Family groups 135
Federal Republic of Germany 39
Fever 74
Fewtrell 5
Fishers exact test 145, 146
Fleisher 6, 95
Flocculants/coagulants 191
Flocculation 193
Flowing stream 184
Flu/cold symptoms 98
Follicular dermatitis 73
Food poisoning 73
Foreshore 53
France 96, 140
Fresh recreational waters 136
Freshwater epidemiological studies 80

G

Gangrene 77
Gastroenteritis 73, 78, 115
Gastrointestinal (GI) ailments 114
Gastrointestinal tract 98
Gelatin capsules 112
Gene probes 80
General Environmental 58
General Environmental Duty 57
General Recreational Duty 57, 65, 69
Genetic structure 184
Geometric mean 137
Geometric mean water quality 143

Germany 185
GESAMP 34, 147
GI illness 120
Giardia lamblia 75
Giardia spp. 73, 75
Good Beach Guide 34
Grantham 5
Great Yarmouth 5, 33
Green Party. *See* The Green Party
Greenpeace 203
Grit 30
Grossly polluted water 94
Guideline standards 27, 80, 134
Guildford 6, 89

H

Haemorrhagic conjunctivitis 78
Hair 91
Halophilic vibrios 94
Health hazards 94
Health risks 5, 71, 89
Hepatitis A 147
Hepatotoxins 79
Her Majesty's Inspectorate of Pollution 31
Highly Credible Gastrointestinal symptoms 135
Highly Credible GI 114, 124
Highly Credible symptoms 120
Holland 185
Hong Kong 96, 140
House of Commons Environment Committee 5, 64, 71, 95
Howarth 5
Human deaths 79
Human embryonic fibroblast (HEF) 161
Human host 91
Human viruses 7
Hydraulic load 190
Hydraulic regimes 107
Hypochlorination 197
Hypochlorite ions 178
Hypochlorite solutions 192

I

Identification 65

Immunity 114
Immunocompromised 74
Imperative standard 134
Implementation of the Directive 56
Inactivation of infectious
    microorganisms 157
Incubation conditions 108
Indicator organisms 29, 79
Industrial effluent 71
Industrial organics 191
Infectious hepatitis 78
Influenza virus 181
Intensive sampling 98
Inter-laboratory precision 111
Inter-laboratory trials 111
Israel 96, 140

J
Jones 5, 6

K
Kasaugamycin 184
Kay 6
Kent 97

L
L. canicola 75, 76
L. hardjo 76
L. icterohaemorrhagiae 75
L. pomona 75
Lake bathing 77
Lake Michigan 129
Lake Ponchartrain 7, 113, 117, 120
Lakes 74
Lamp modules (UV) 194
Lampeter 5, 71, 129
Lamps 189
Langland Bay Study 16, 37, 98, 141
Lateral mixing 190
Legal Issues 5, 49
Legionella pneumophila 181
Legionella spp. 73
Legionellosis 73
Leptospira spp. 73, 75, 149
Leptospirosis 73, 94
Lethality 178

Lightfoot 95, 139
Likelihood ratio x^2 121
Linear regression 7, 116, 119
Liquid gas chlorination 197
Liverpool Crown Court 69
Local authority 5, 33
Local variation 115
Log odds of disease 120
Log odds of gastroenteritis 7
Log_{10} standard deviation 106
Logarithms 106
Logistic regression 7, 120
London 6, 175
Lord Crickhowell 17
Lyme Regis 98

M
Male-specific bacterial virus 161
Mamaroneck 131
Mandatory standards 27, 80
Maryland 6, 157
Measurement error 139
Medical and clinical follow-up
    examinations 135
Medical Research Council 95
Medmenham 6
Member States 20
MEMBIO™ 161, 164
Memtec America Inc. 6, 157, 161
Memtec CMF 159, 161
Meningitis 73, 78
Meningoencephalitis 73, 76
Methodological weaknesses 114
Microbial populations 6
Microbial standards 30
Microbiological AQC 111
Micrococcaceae 91
Microcystins 79
Microcystis aeruginosa 79
Milk 74
Mineral oils 20, 29
Minimum infective dose 74, 78
Minister of Agriculture, Fisheries and
    Food 16

Ministry of Agriculture, Fisheries and Food   26
Minnesota   193
Minor ailments   204
Miscounting and visual fatigue   109
Missing data points   118
Mississippi river   77
Mono-clonal antibodies   80
Moreton   97, 141, 147
Moulds   181
Mountain streams   74
Multiple-tube dilution counts   105
Municipal Waste Water Directive   38, 40, 41, 43, 44
Mutagenic compounds   172
Mutagenic effects   186
Mutation   183
Mutations in human cells   178
Mycobacterium tuberculosis   181
Myocarditis   78

N

N. fowleri   76
Naegleria spp.   73, 76
Nasal passages   76
National Rivers Authority   25, 38, 57, 66, 203
National Technical Advisory Committee   132
Natural breaks in measured indicator organism density   114
Natural enrichment   55
Nature Conservancy Council   56
Nausea   74
Negative charge   178
Neurotoxins   79
New Jersey Health Department   139
New Rochelle   131
New South Wales   160
New microbial strains   184
New York City   6, 7, 113, 117, 120
Nicrococcus radiodurans   181
NIMBY syndrome   43
Non A non B hepatitis   94
Non-aromatic amino acids   178

Non-pathogenic indicator bacteria   79
Non-swimming control groups   133
North America   194
North American standards   134
North of England   37
North Sea Conference   30
Northwest Bergen County Utilities Authority   194
Norwalk virus   78
Notifiable disease   75
Novobiocin   184
NRA   5. *See also* National Rivers Authority
Nucleic acid   175, 176, 184, 186
Nutrient loads   191

O

Offences of polluting controlled waters   61
OFWAT   17
Ohio River   90, 129, 133
Ohio River-Pool   131
Oldridge   5
Olfactory tissue   76
Olivieri   6
Ontario   6, 175
Organs and orifices   91
Other recreational water exposure   143
Otitis externa   73, 77
Ova   147
Ozonation   42, 197

P

Paignton   98
PAM. *See* meningoencephalitis
Paralysis   78
Parasite   147
Paratyphoid   73
Pathogens   7, 72, 80, 198
Pearson test   122
Penicillium roqueforti   181
Peptone water   110
Perceived symptom attack rates   143
Percentile   106
Perception data   139
Person to person spread   75

Subject Index

pH 20, 21, 29
Pharyngo-conjunctival fever 94
Phenols 20
Phenomena 184
Phosphate control 192
Photochemical changes 175
Photoreactivation 182
Pike 6, 203
Plastic litter 148
Plastics 30
Plausibility 95
Plesiomonas spp. 73
Pleurodynia 78
Pneumonia 79
Poissonian theory 110
Polio 178
Poliomyelitis 132
Poliovirus 78, 181
Politicians 206
Pollution of Beaches 64
Ponds 74
Portsmouth 141
Post-sedimentation filters 191
Poultry 74
Predation 184
Prestatyn 98
Primary effluent 183
Probability of disease 120
Probability of gastroenteritis 7
Prospective epidemiological studies 135
Prospective methodology 131
Protozoa 181
Protozoan cysts 157
Protozoan parasites 74
Pseudomonas aeruginosa 73, 91, 92, 98, 146, 181, 185
Public concern 129
Public health 52
Public Health Laboratory Service 6, 36, 89, 94, 111
Public health risk 149
Public Perceptions 42
Pyrimidine(6-4)pyrimidone photoadduct 178

Q

Quality control charts 111
Quality standards 54
Quantal assays 161
Quartz envelope (UV) 194
Questionnaire 95

R

R^2 value 115
Ramsgate 97, 98
Rash 78
Rate difference 115
Rate difference of GI symptoms 116
Raw seafood 77
Reasoned opinion 22
Receiving waters 198
Recent epidemiological investigations 137
Recent water quality standards 135
Recording of duration of exposure 141
Recreational Duties 58
Recreational water 71
Recreational waters 6, 89, 94
Reference group 114
Regression analysis 7
Regression coefficient 122, 123
Regression equation 115
Regression models 149
Relative measures of effect 115
Relatively unpolluted beaches 135
Replicate determinations 109
Respiratory disease 78
Respiratory infections 148
Respiratory symptoms 98
Restrict public bathing 53
Retrospective research design 131
Ripa di Meana 65, 69
Risk to the health of bathers 23
RNA 176, 180
Robens Institute 36
Rossi 5
Rotavirus 78, 148
Round Corner treatment plant 160
Route of spread 93
Royal College of Physicians 141

Royal Commission on Environmental Pollution 36, 41, 54
Royal Marine Depot at Walmer in Kent 90

S

S. boydii 76
S. dysenteriae 76
S. flexneri 76
S. sonnei 76
Saccharomyces cerevisiae 181
Salmonella enteritidis 181
Salmonella paratyphi 181
Salmonella spp. 73, 90, 143, 147, 157, 185
Salmonella typhimurium 181
Sampling 20, 55
Sampling - method and position of sampling 107
Sampling and analysis 106
Sampling regimes 137
Sarcina spp. 91
Scarborough 34, 36
Scientifically robust epidemiological research 139
Scotland 94
Screens 30
Sea Fisheries Committees 56
Sea water 147
Sebaceous glands 91
Secondary effluent 183
Secondary treatment 42
Secretary of State for the Environment 25, 60
Sedimentation 164, 184, 191
Selectivity 108
Self-aggregating microbial surfaces 191
Sensitivity of microbes to UV 181
Sewage 78
Sewage effluent 62
Seyfried 96, 116
Shear forces 191
Sheep 79
Shell fisheries 42
Shellfish Waters Directive 55

Shigella dysenteriae 181
Shigella spp. 73, 76, 185
Shigellosis 77, 94
Sidmouth, Devon 108
Single European Act (1987) 52
Site surveys 106
Skin 91
Sludge 55
Social background 143
Sodium azide 108
Sodium bisulfite dechlorination 197
Sodium lauryl sulphate 108
Soft tissue infections 73
Solar UV 183, 184
South Wales 97
Southsea 98, 141
Spain 96, 140
Specificity of association 95
Sports Council 58
Sports Council for Wales 58
SS 160
St Luke's Hospital 6
Standard deviation 108
Standards 28, 80, 129
Staphylococcus aureus 91, 146, 181
Staphylococcus spp. 73, 98
Statistical aspects 6, 105
Statutory Water Quality Objectives 25, 203
Sterilization 108
Stevenson 131
Stomach pains 74
Storms 107
Stormwater overflows 31
Strength of association 95
Streptococcus bovis 79
Streptococcus equinus 79
Streptococcus faecalis 146
Streptococcus pyogenes 181
Streptococcus spp. *146*
Stuttgart disease. *See L. canicola*
Sulphur dioxide dechlorination 197
Sunwater Limited 197
Surface active substances 20
Susceptible host 94

Suspended solids 189
Swansea 141, 142
Swimming 77
Swimming pool 90, 94
Swineherd's disease. *See L. pomona*
Switzerland 185
Synergistic bacteria 109

T

Tacoma Park Pool 129
Tarry residues 29
Teepol broth 110
Temporality 95
Terminal disinfection 6
Territorial waters 59
The Green Party 35
Thermal spring water 76
Thermo-tolerant coliform bacteria 108
Third North Sea Conference 16
Thymine 178
Tidal and diurnal effects 107
Timonium 6
Toronto Health Department 134
Total coliforms 90, 161
Total GI 114
Toxins 79
Trans-membrane pressure (TMP) 159
Transparency 20, 27
Transport and storage of samples 107
Travel history 143
Trichobilharzia ocellata 94
Trickling filter plants 164
Trihalomethanes (THM) 192
Trojan Technologies 6, 175, 193
Tryptophan amino acid 183
Tryptophan independence 183
Typhoid 73
Typhoid fever 90, 94

U

UK recreational water 7
Ultrasonics 80
Ultraviolet 107, 197
Ultraviolet (UV) light 175
Ultraviolet treatment 42
Uniform Fire Code 192, 193

United Kingdom 56, 96, 159
United Kingdom Government 41
United Nations Environment
  Programme 33, 147
United States 75, 159, 168
United States Environmental Protection
  Agency 7, 90, 125, 194
University of Wales 129
Upland lakes 93
Urban Waste Water Directive 30
US federal bacteriological water quality
  standards 6, 113
US Public Health Service 95
USA 140
USEPA. *See* United States Environmental
  Protection Agency
USEPA guidelines 125
UV absoring colloids 192
UV disinfection 6, 175, 194, 198
UV radiation 30, 187
UV reactor 189
UV transmitting properties 189

V

Vcrsorgeprinzip 39
Vibrio 149
Vibrio cholerae 01 77
Vibrio comma 181
Vibrio parahaemolyticus 77
Vibrio spp. 73, 77
Vibrio wound infection 77
Victorian heritage 33
Vincent 5
Viral infections 78
Viral pathogens 148, 157
Viruses 148, 157, 181
Volunteer populations 141
Vomiting 74

W

Waldwick, New Jersey 197
Wales 5, 6, 56, 94, 130
Washington 193
Waste water 187, 190, 198
Water Act 25
Water Act 1989 63. *See also* Water Act

Water analysis 98
Water and Sewerage (Conservation, Access and Recreation)(Code of practice) Order (1989) 59
Water contact sports 42
Water quality classifications 60
Water Quality Objectives 41
Water related leptospiral infection 76
Water Research Centre 6, 42, 105
Water-borne infections 71
Weil's disease. *See L. icterohaemorrhagiae*
Wet hair 135
White Paper on the Environment 41
WHO 97
Willinghan 6
Wirral 141
Wound infections 73, 94
Wyer 6

Y
Yeasts 181
Yersinia spp. 73
Yorkshire 75
Yorkshire and Humberside Pollution Advisory Council 37

Z
Zoonotic 73, 79